My Inventions For A Horse

By

Wagdy A. Assawah

Bloomington, IN  Milton Keynes, UK

authorHOUSE

AuthorHouse™
1663 Liberty Drive, Suite 200
Bloomington, IN 47403
www.authorhouse.com
Phone: 1-800-839-8640

AuthorHouse™ UK Ltd.
500 Avebury Boulevard
Central Milton Keynes, MK9 2BE
www.authorhouse.co.uk
Phone: 08001974150

First published by AuthorHouse 4/27/2006

ISBN: 1-4184-1186-8 (e)
ISBN: 1-4184-1185-X (sc)
ISBN: 1-4208-5245-0 (dj)

Library of Congress Control Number: 2003098543

Printed in the United States of America
Bloomington, Indiana

This book is printed on acid-free paper.

TABLE OF CONTENTS

FOREWORD

Many people have made this book possible, but I should like to single out Mr. Nader Assawah, my son, who took part as a co-inventor and was my financial supporter to secure a patent on our invention entitled, "Manually Driven Water Propulsion Device". It was he who first suggested to me to document the whole gamut of my inventions, in which I have spent more than fifteen years thinking, perfecting, drawing, and writing, lest they get lost to oblivion or worse, plagiarized. Elated by my approval to publish and let not the work perish, he was fueled to design and execute the book cover. He intuitively used Benjamin Franklin's kite-and-key experiment to symbolize my humble efforts.

Whether I have succeeded in making my inventions or ideas mainstream is not for me to say, but Mrs. Rhoda Okaily has been tireless in her encouragement to me and in her hard work on behalf of the book.

A number of friends have also guided me through different complexities: Dr. Aly Okaily, principal engineer, Canberra Industries, Inc., and Mr. Wagdy Okaily, clinical research associate, Parexel.

I am also most grateful to my wife for her efforts in tabulating my old manuscripts and for her continued support during taking over this task.

Finally, I would like to express my deep thanks to my daughter-in-law, Kimberly Selmar Assawah, for putting up cheerfully with the changes that sometimes entailed the scrapping of a whole day's typing. Also credits are accorded to Mrs. Beth-Ann McHugh for putting the work in its approbatory layout.

That the book has appeared at all is due to their hard work. However, the whole responsibility for the book rests within me, since the ultimate decision as to what went in and what was omitted is mine.

Wagdy A. Assawah

PROLOGUE

The United States of America is the only country in the world to which one can come and declare it to be his own homeland. The awesome ceremony of taking the oath of allegiance was to me a moving experience and a soul impelling commitment for doing my best in serving my new country and to pursue my happiness and the happiness of others too.

Unfortunately I was not given the chance to oblige in spite of my outstanding credentials and my honorable history in my old country, Egypt. In my new country, U.S.A., I worked as a substitute teacher at several middle and high schools. And even the opportunities to be a "glorified babysitter" were surprisingly coincidental with either sizzling hot days or snowy, stormy conditions; yet with plenty of leisure time on my hands, I strived, though from afar, to make a difference. For that, I thought of several inventions that could add to our national prosperity and to promote life for many human beings cohabiting our mother earth.

Many people are born with innate abilities to excel in one field or another. So we find painters, musicians, writers, scientists, daredevils, etc. Finding that I enjoy the ability to identify problems and to see solutions for them, I decided to use my God-given gift for the betterment of mankind. Hence, I took to the drawing table and made its paperboards my canvas, first to draw my visualized contrivances, and second to preserve them from disappearing into the wilderness. The grandeur of our U.S.A. culminates in that no man faires above another, and that we are all born with inalienable rights to pursue happiness. Also in this land of the free, our Constitution gives every man the right to voice his opinions without repercussions. Henceforth I believe that every man has also the right, if not the obligation, to document his mentations and leave them for others to evaluate. So writing down my inventions and my opinions constituted a humble effort that I dutifully feel obligated to leave behind. As to how valuable they are, it is not for me to judge.

Hopefully, future generations will find in them ideas that can be made use of; blueprints of apparatuses that can be constructed and metamorphosed for the better most. At this point of my transcribing, it may be worth mentioning that I neither claim ingenuity nor do I seek wealth, and that is why I entitled my book, *My Inventions for a Horse*.

What I am seeking is self-gratification. And if only one of my inventions gains acceptance, or one of my ideas proved useful, this will justify the existence of the book and the existence of its author. And, for lack of better words, let me quote from President Theodore Roosevelt verses mirroring what reverberates inside me.

Who knows the great enthusiasm
The great devotion
Who, at best, knows in the end
The triumph of high achievement,
And who, at the worse, if he fails.
At least fails while daring greatly,
So that his peace shall never be with those timid
Souls who knew neither victory
Nor defeat.

COMMEMORATION

In Remembrance of My First Mentor:

MY FATHER

His outstanding advice to me was:

A tentative translation of this Arabic aphorism will be:

"No one merits his birth if he solely lives to fulfill his own felicity."

My father's guiding byword became my motto in life, and became an indicative to measure up to. He also taught me the sanctity of books. And that whenever I pick one, I should ritually greet with a gentle puff to rid it from the dust clinging to its cover. Though a modest gesture, yet it went a long way to enforce in me the value of past achievements coming our way from those authors who came before us. Nevertheless, he cautioned me from falling for the luster of printed words, yet per contra, to diligently question them.

DEDICATION[1]

TO ETIMAD (Amy)

---

[1] *Mr. Andy Rooney looked upon the nearly blank dedication pages showing a single name, as prodigal paraphernalia. The fact is, dedication pages are well-spent pages. Besides recognizing support and help from the dedicatees, they represent a noble passion worth keeping verdant.*

LETTRES MISSIVE

To

Astronomers	**Biologists**
Encyclopedists	**Environmentalists**
Geophysicists	**Inventors**
Oceanographers	**Politicians**
Physicians	**Theologians**

Researchers in every Scientific Endeavor

And for my shortcomings, I say:

"If perfection means acceptance, I'd rather embrace a measure of imperfectability so others continue to chase impeccability"

Wagdy A. Assawah

INTRODUCTION
THE PASSIONATE JOURNEY OF
INVENTIVENESS

Through the ages, man used his ingenuity for making inventions to upgrade his living conditions, to appease the gods, and to improve his environmental conditions. In the early epochs, inventiveness was sluggish, and it is now in full swing. This is due to population surges, and with it inventive minds, the advancements in different scientific disciplines, the availability of sophisticated technologies, and the great ease with which we can communicate with each other. However, inventors are like philosophers, social reformers, scientists, painters…etc., in being ahead of their time. Their futuristic visualizations, intuitive concepts, and conjectured inventions, targeting the prosperity of their societies, are bound to be misunderstood. That is because their contemporaries are neither ready to accept new ideas nor to allow for them to sink in. However, the recognition of the truly gifted is inevitable by their successors.

Taking interest in publishing my own inventions, I thought at first that it may be satisfactory to confine my efforts to organize them in a chronological cavalcade. Yet for the sake of clarity, it was necessary to give an account of my reasons for disclosing my work and to justify the choice of *"My Inventions for a Horse"* for a title.

Spending over fifteen of my cognitive years doing few inventions, having several encounters with the Patent and Trademark Office (P.T.O.), with companies promoting inventions and brokers negotiating contracts with the industry, endowed in me a measure of proficiency. Also as a byproduct of gained experiences and other assimilated practices, I developed certain convictions and ideas. I was then motivated and consciously compelled to write about my views concerning inventiveness, inventors, and inventions. Though my work may not gain unanimous acceptance, I may be credited for arousing interest in certain controversial aspects and in pointing out different problems of this life-changing endeavor. But for now, let me spare the interesting to the last. Meanwhile, let me reveal from experience what is taking place in the passionate journey of inventiveness.

Wonder not, my reader! The invention endeavor is: figuratively speaking: a beast at large, though not of its own making. And that is why the more the reason it should be subdued for the benefit of our nation in particular and humanity in general. Untamed, this beast will be like the ever growing ozone hole. Both are damaging and we cannot afford to have a passive stance. The first will harm the new developing industries, the spine of our prosperity, and the second will inflict debilitating deformities in the populous. For our lurking beast, we need regulatory laws to put an end to the exploitation of inventors, the bilking of industry, and to the unforeseen harms to the nation. In this business, there is no shopping around. In the so-called open market, the exploitation of inventors is a steadfast ritual and the greed in levied fees is kept on a par by corporations and brokers all alike. Hence, the inventor is left to choose between either the inferno or brimstone.

In this waged war, the inventor will be fighting those who mobilize the beast: i.e., **"The Corporation Services"** and **"The Brokers."** The inventor at one time will be an easy prey for the corporations responsible for submitting his work to the P.T.O., and at another, he will be at the mercy of one broker or another, acting as a liaison officer for securing a license with the industry.

The corporations will give the inventor the cold shoulder till they get their cold cash and get it in plenty. And take it from me, "Taking your patented invention to the Market" is a hollow promise intended to be disregarded. It is only a slogan invented by the C.E.O.s of those corporation services for taking the inventor to a ride. A living proof of the exploitation made by these corporations engaged in collecting unfair fees or for using deceptive practices was the court action brought against a well-known corporation in 1998. Pursuant to the court action was the establishment of a redress fund to pay back a pro-rata share for the claimants.

To make sure that his invention is not going to be neglected, the inventor borrows the needed sums that satisfy the ravenous appetite of the corporation. Yet such payments will only get him, after nearly three years of cross examinations of his work and myriads of costly communications, a patent on an evanescent paper that can be pulverized into nothingness by the broker. Meanwhile, the inventor will be left behind, drowning in a sea of debt.

With a patent granted, comes the turn of the broker who wants his voluptuous cut. Brokers habitually wait in the wings for the patent to be awarded to make the kill and to enjoy scavenging on a free game. At first, a broker, though warranting no success, asks for an avaricious fee of $9,000.00: allegedly for negotiating a license with the industry over a six-month period. And if success emanates, he will be eligible for a lion's share amounting to 35% to 40% in royalty.

Now, if the inventor succumbs to the exorbitant wages, he will be looking with lamentation at his efforts going to the unworthy. The inventor has endured several scrutinizing examinations for his submitted work; he has been jumping financial hurdles and he has been working around the clock for several years to achieve a tenable result. And now for someone to come out of the blue levying austere demands, with no commitments, will be totally unfair and unethical.

In fact, those brokers have forced themselves on the stage and are driving a wedge between the industry and the inventors. Putting on monk's cloaks, they convince the industry that dealing with inventors is a bad idea because inventors—so they say—are litigious and greedy. Meanwhile, they convince the inventors that they are the defenders of the right and that they will fight the beguiled industry to get the inventor his rightfully-earned royalty.

Unfortunately, the end result of this perilous journey seems as if it is a dilemma of one unlucky inventor. But a careful analysis will reveal several malefic states of affair. The broker, apart from fitting the inventor with a shaft, they are also throwing a monkey wrench in the nation's industrial future. Unaware that they are losing valuable opportunities, the industry will continue, unconsciously happy with their humdrum activities. Losing interest in new products, they will survive as long as their assembly lines continue to produce the usual products offering the blissful status quo.

Consequently, the biggest damage will befall our national industries. It is the combined unwelcomed roles played by the corporations and the brokers on one hand and the tactless authorities that cannot differentiate between the momentous from the trivial on the other hand. In the long run, neglecting promising inventions will yield harmful effects: Namely, the deprivation of the nation from having useful new products, the loss of needed jobs for the working force, and last but not least, the loss of revenues to the treasury from selling new products locally and universally. And because one thing leads to another, joblessness and loss of revenues will be undoubtedly contributing factors in setting the stage for dreadful recessions.

To Disclose Or Not To Disclose:

At first, I was reluctant to collect and collate my inventions in a book form. Motivated by the news that 700 inventors from 44 countries went to Geneva in April 2000 to show off their creations, and that more than 65,000 visitors from all walks of life were attracted to this International Exhibition, made up my mind to venture the trial, feeling that if I resign to do nothing, I'll be betraying myself, and others too. Apart from being encouraged by the Geneva Inventions Exhibition, there were other compelling reasons to do what I felt was imperative.

Men Have The Seeds Of Death In Them:

- Plagued like all living creatures by mortality, and neither being a genetically altered author, nor an interphase of man and machine that may give me a longer life span over that of a septagenerian approaching quickly his eighties, I chose to invest my inventive exertions into a compendium. That is because compendia can outlive their authors and can act as time capsules trusting the work to a fertile soil to take hold and to bear fruit.

- Also I owe it to my integrated knowledge and to my diversified experiences over 55 years in inter-disciplinary scientific endeavors to own what I did. This is specially so, since what I have executed is documented and dated by governmental institutions. Thence, I decided neither to leave it for hijackers to scavenge on a feat of forgotten inventions nor to let my rooster advertise the dawn on another man's roof.

- It is also said that ingenuity is one percent inspiration and ninety nine percent perspiration. By the same token, inventions withheld from sight will be of ulterior value while publicizing allow their availability to be made use of, corroborating the adage, "You own only what you give".

- Unrealized projects are facts of life, so I decided not to leave my ideas to go to oblivion. I fought to make them and I must fight to keep them under the limelight.

- It was difficult to keep my inventions bottled up. And now with their accessibility, they will unleash the power of imagination in younger inventors. They may be forerunners catalyzing chain reactions of new inventions. In this age of satellites, Internet, and global information, one beneficial apparatus will be half a world away in no time.

Redding my inventions in a book form was an arduous task. Approaching my eighties with blurred vision, a mutinous heart due to my mitral valve regurgitation, rickety hands, and a forgetful mind; made the undertaking an uphill battle. Yet, in spite of my body's incompetencies, it was forced to do the bidding of the soul. This was a mission I took issue with because it was uplifting to me. And to those who volunteered with obstacles to hamper my bear, I say thanks but no thanks. And for a deductive syllogism, the aroma of the incense will not be perceived till tested by fire.

Why *My Inventions for a Horse*

No one can complain over the reasonable fees asked for by the P.T.O. concerning filing, issue, maintenance, and other required fees for specific purposes. The problem lies within the majority of the mediators, i.e., the Corporations and the Brokers. That is because these mediators are spoiling the healthy climate between the inventors and the industry on one hand, and the industry and the

9

national interest on the other. The exorbitant fees they ask for are no less than ransoms levied by a rowdy grinch.

Unfortunately, this ugly extortion is running amuck under the noses of the concerned authorities. Leaving this exploitation perversive and pervasive means that we are allowing the rules of the jungle to prevail and that we are drudged to "Invention Incivility." The fact is, inventions neither grow on trees nor are they plucked from thin air. Each invention is the product of toil over two to three years of careful planning and designing by a devoted inventor. Being the mediators' cash cows, inventors neither deserve to be scapegoats sacrificed on the altar of their greed, nor to be anticipated as the golden geese to be slain for their golden eggs.

From the earlier mentioned perspicuous exploitations, I'd rather publish my inventions in a book form than to succumb to the avaricious demands of the mediators. Hence, *My Inventions for a Horse* was an appropriated title for the book. Thereby, with the blissfulness of delivering my humble ideas and my innovative contrivances to a wider populace, I dare-say that I savor the joyance of spending my life for a worthy cause.

PART I

INVENTIVENESS A REVOLUTIONARY STRIVE AND AN EVOLUTIONARY ENACTMENT

Scientific knowledge has turned out to be conjectural, permanently open to revision in the light of experience.

Karl Popper

CHAPTER 1

INVENTIVENESS THROUGH THE AGES

Our world was subjected to sways of innovative revelations. The accepted views recognize two great ones, namely: "The Great Agricultural Revolution" and "The Great Technological Revolution" contributed by our species, *Homo sapiens sapiens.* However, our nomadic hominid ancestors *Homo faber,* i.e., the tool maker, contributed to inventiveness in their own modest way. Yet their humble efforts were excluded from the picture because they belonged to a very distant past measured by millions of years and they were bipedal primates, and not mankind.

THE HOMINIDS' WAY OF LIFE

Adverse conditions taught the nomadic hominids existing 2.5 million years ago to maintain their livelihood by gathering fruits and by scavenging on dead kills made by other beasts. Going hunting, they used chipped flints as tools for the purpose. From brush fires they learned the value of flames in warding off wild beasts. They also stumbled over the benefit of eating tender meat from animals caught in the inferno. Enticed by the phenomenon, they tried to create their own fires. They found that by striking flints against each other, they can ignite fire in dry weeds. The use of fire was thus introduced as early as 1.5 million years ago.

Woven objects made by hominids from animal hair were also found in archaeological records. This proved that weaving to make blankets and the like with the purpose of providing warmth predated the birth of agriculture. The extraction of bone marrow from carcasses, the chipped flints, the making of fire, and the weaving of objects from animal hair may be looked at as modest achievements. Yet to our hominid ancestors, these simple tools used during this long period of their evolution, gave them the needed competitive edge to overcome the unmatched strength of contemporary beasts and to endure the hostile environmental condition. Thus preparing

13

the stage for the emergence of *Homo sapiens sapiens,* our own species, who developed from maundering savages to sophisticated technicians.

THE AGRICULTURAL REVOLUTION

It is also called "The First Great Technological Revolution." This revolution started about 10,000 years ago around the fertile valleys of North Africa, Mesopotamia, and Asia Minor. During this revolution, *Homo sapiens sapiens* gave up the nomadic style of life and took to living in cities around river valleys. Dwelling in cities necessitated the presence of social institutions for keeping law and order. It also encouraged developments in scientific and artistic endeavors. In this era, man knew how to domesticate animals, thus learning about animal husbandry. He also learned how to plant, grow, and harvest crops to sustain his existence. Fearing drought and shortage of food, he learned how to store crops from superfluous years to lean years. This necessitated how to keep records and develop the mathematics needed for doing the job.

Living in agricultural communities, man came to know about the seasons, to know about the suitable times for either sowing his crops or reaping his harvests. He also developed technological skills for digging irrigation canals and for making bridges to cross them. He also developed architectural abilities to build homes to live in, and skills to embalm the dead and to make tombs for their corpuses.

Farming has become more and more scientific. The open field system became a common place, while fencing was for sheep rearing, cattle breeding, and pig keeping. Poultry and calves on the other hand were often raised indoors as if in a factory. After discovering the "New World," plants such as tomatoes and potatoes were introduced to Europe and to other agricultural countries in Africa and Asia. In the late 1800s and early 1900s, threshing and reaping machines came into use. Fertilizers, pesticides, new plant strains able to resist disease and to yield improved crops were a common practice. This allowed the rich countries to produce more food than they need and to export the excess to poorer countries.

THE INDUSTRIAL REVOLUTION

It is also called "The Second Great Technological Revolution." This revolution occurred centuries later from the Agricultural Revolution. It overlapped with it and started about 250 years ago.

In it, the muscle power of animals was replaced with coal-fired steam engines and later, about one hundred years ago, these were also replaced by gasoline driven internal combustion engines. Damming rivers and building hydroelectric turbines worked by the force of falling water made electricity available. Being a flexible source of power easy to send through cables, it was preferred over the two earlier sources of energy. Many people moved from the countryside and began to work in mining and in factories. This, in its turn, encouraged the development of transportation systems to bring raw materials to the factory and to take the finished product to outlets for distribution. So railroads, highways, and water canals were as essential as the factories themselves.

In this Industrial Revolution, our classical gadgetry began to change in leaps and bounds. This acceleration in the tempo of inventiveness taking place in the last few decades was due to several factors: The exponential increase in populations and consequently the increase in inventive individuals. The ripple effect of our increasing knowledge compared to that of earlier generations, and the urge to perfect everything from the simplest to the most sophisticated.

RECOGNIZING CORRELATED INVENTION
CONGLOMERATES

In the following, though not exhaustive, is a modest sampler with the purpose of showing how our earlier inventors in different endeavors, attempted to procreate second, third, and later generations. This resulted in recognizing conglomerates of inventions stemming from a common necessity. Thence, the discussed herein, is with the purpose of outlining such conglomerates and the contained families within each conglomerate and the contained lineages within each family.

On The Agricultural Front:

We went from using plows that has been pulled by beasts of burden for centuries to the use of powerful mechanical plows able to break up several rows and to sow them at the same time. For cutting ripened grains, we went from using a sickle to use a reaping machine, the prototype of which was invented by Cyrus Hall McCormick in 1839. And for separating the chaff from grain, we went from using a winnower to use a threshing machine.

Wagdy A. Assawah
On the Home Front:

For lighting our homes at night, we went from bioluminescence produced by firebugs entrapped in glass bottles emitting their soft light by oxidation of luciferin to candles to lamps with wicks for burning oil, to electric bulbs invented by Thomas Edison to fluorescent lamps invented by the American physicist Arthur H. Compton in 1934.

For washing, we went from the corrugated board to the programmed washing machine. And from the hand wringer to the automatic drier that shuts off when laundering is done. For cooking, we went from the stove to the electric ring to the microwave. For sweeping the floor, we went from the broom to the electrostatic sweeper to the vacuum cleaners. For knowing time, we went from the first water clock invented in 159 B.C. by Clepsydra: a Roman prostitute who wanted to give equal times to her patrons: to the sundial to the grandpa standing clock to the classical wristwatch to the digital watch. And for the record, it was the Swiss Physicist, Carl August Steinheil who built in 1839 the first electric watch.

To open closed doors, we went from unfastening a latch to twisting a key in a keyhole, to slide a magnetic card in a slit, to show our fingerprints on an electronic plate. For sleeping, we went from a hay-rick to cotton padded mattresses to spring mattresses to waterbeds to air-beds.

On Cloth Making:

We went from the hand driven spinning wheel for making a single thread to "Spinning Jenny" with multiple spindles, to weaving machines using shuttles for interlacing threads into woven fabrics. Latter, James Watt and Matthew Boulton in 1785 installed a steam engine with a rotary motion in a cotton-spinning factory at Pappelwick, Nottinghamshire. These machines altered work methods that had not changed for hundreds of years. And it was Isaac Singer who invented the first successful sewing machine in 1846. For fastening our clothes, we went from ribbons to safety pins to buttons to slide fasteners commonly known as zippers, invented by Swedish inventor, Gideon Sundback in 1912, to Velcro of small hooks that stick to corresponding small loops.

We use different needles for different jobs: Straight steel-eyed needles for threads to mend torn clothes or for surgical sutures to sew parts of the living body. Curved needles for cobblers to mend shoes or for sewing leather goods. We also use needles in household sewing machines to make garments, and we use circular sets of needles in mass production machines for weaving socks or for weaving flannel underwear. We have also eyeless needles e.g. long straight ones for knitting pullovers and the like and hooked ones for crocheting.

On Writing, Printing, and Photocopying:

The first writing using picture-signs was invented more than 5000 years ago. And when man needed to record his transactions, he used bones, animal leather and later fired clay tablets to write on. He also used walls of temples to glorify the triumph in battle over his enemies and to document the different religious rituals practiced to appease his deities. Yet the requirement for writing down his experiences necessitated a newie medium. For this purpose, the Ancient Egyptians made a kind of paper from papyrus, a plant that grows in swamps in Egypt. However, it was the Chinese who invented paper as we know it today. About 1,900 years ago, they learned to separate fibers from mulberry bark. They soaked them and then dried them, making a flat, dry sheet that they could write on. Paper is still made of plant fibers. Some of the best is made from cotton, reminiscent of parchment that was prepared for writing from the skin of sheep.

Writing on a new medium, i.e. paper, called for the invention of ink and necessitated the use of dip-pens. Pencils came also into use. Each contained a slender cylinder of a solid marking substance called graphite. This latter substance was a French invention made from a soft black form of carbon. In 1780, Scheller constructed the first fountain-pen. It contain a reservoir that automatically feeds the writing point with ink. For a long time, fountain-pens were favored by many. However, pilots of W.W. II found that the pens leaked at higher altitudes. Looking for a better alternative, they find in the ball-point pens their heart's desire. These were invented by the Hungarian Lajos Biro in 1938. Each has a small rotating ball that inks itself by contact with an inner magazine.

Like almost every invention, paper, pens, and pencils stimulated a virtuous cycle of "invention-helpers." Of the paraphernalia servicing paper one may mention, the folders, paper-clips, hole-punchers, staplers, numbering machines, paper-cutters, adhesive tapes, rubber stamps, paper board…etc. As for the accessories invented to enhance the efficacy of pens, one may stress the value of invented inks that came in different colors, the porcelain ink-pots, the spongy blotting paper held by a blotter, the pen holders, the correction fluids…etc. Concerning the corollary gadgetries that ensued the pencil, came the pastel color-sticks, a rainbow of colored lead-pencils, erasers, an assortment of pencil-sharpeners, pattern-templates…etc.

Concerning printing, the books of yore were copied by hand. The earliest printing, using wooden blocks, was done in China as early as A.D. 500. Johannes Gutenberg of Germany founded modern printing with moveable type in the 1400's. The invention of printing made books available to everyone who could read. Now there are different methods of printing. The most common method is called letter-press printing, in which the ink is put onto raised images and letters. In intaglio or gravure printing, the image is not raised, but etched, while offset lithography uses printing plates that are made photographically.

For typing, we used typewriters. In the early machines, e.g. "Hammond's Typewriter" of 1880, the hammer hits the back of a piece of paper, pressing it against a letter on a fixed cylinder. On the other hand, typewriters of the early twenty century, were operated by means of a keyboard, and the deployed types strike a ribbon to transfer carbon impressions onto the paper. Nowadays, most of our typing is made on computers, and the displayed text can be retrieved from a printer hooked to the computer. However, our competitive edge technologies allow nowadays a user to dictate what he wants to inscribe. And as if by a magic wand, the programmed computer gets it on paper.

For making a legible copy, we went from a duplicate made by carbon paper to copies from a congealed jell-copier to the black-and-white copier invented by Chester Carlson in 1930. Carlson had invented a dry process based on the ability of an electro-statically charged plate to attract powder in the image of the original text. Now we have modern color copiers and laser copiers that can make unlimited numbers of xerographic duplicates from an original.

The modern methods of xerography condemned to oblivion an earlier instrument invented in 1723 that was used specially for copying maps on a pre-determined scale. This instrument was called the "Phantograph" The ingenuity of those who came before us and their disappearing gadgetries shouldn't be belittled or forgotten. Hopefully someone will address the calling, and charm us with a book on this inscrutable subject.

Wagdy A. Assawah

On Sending Messages:

We went from sending messages on shaven heads of slaves and waiting for their hair to grow to send them to the addressee where the slave can be wasted if the message is of a secretive nature, to relayed smoke signals, to messages on camel or horse backs. In 1792, Claude Chappe, a French man, invented the mechanical semaphore signal. They were mounted on towers in France, Europe, and North Africa.

And we went from the black penny of Queen Victoria in the early fifties of the nineteen-century to organized mail by post offices. And we went from sending telegraphs invented by Wheatstone and W. F. Cooke to faxing to e-mailing.

On Money Exchange:

The earliest form of trade was by bartering. Then other things came into use as means of payments. In the past, people have used shells, beads, salt, grains, and cattle for the purpose. The use of coins was first used in China and ancient Greece. They were made of silver or gold and were stamped by the mark of the government. The stamp show how much each coin was worth. Coins remained popular for centuries. Later, people began to use paper money backed by the government endorsement, or use checks which order a bank to pay money to a bearer as instructed. Traveler checks, drafts purchased from a bank, were also used either for cashing or as a precautionary measure against forgery. Nowadays, credit cards authorizing purchases on credit are now widely used instead of cash. For sending a specified sum of money, we either use money-orders issued by a post office, a bank, or a telegraph office.

Coming into use is the electronic banking, which facilitates transmitting money from place (A) to place (B) in a jiffy.

On Harnessing Energy:

Windmills were first built in Persia in the seventh century. Usually they had four big vertical sails to harness the power of the wind to turn a grindstone or to drive a water pump. The modern counterpart of the windmill drives a generator. To extract as much energy from the wind as possible, the huge rotor blades are kept facing the wind. This is achieved by mounting the turbines on turntables placed at the top of posts carrying them. And providing the turbines with sensors enabling their computers to control the movement of rotors and produce optimum power in all wind conditions.

Waterwheels were invented by the Greeks in the first century. Now we have modern turbines in hydroelectric power stations to produce electricity. Also, in "Nuclear Plants", the energy from the regulated nuclear reactions can be controlled to provide us with steam, which in turn drives turbines that generate electricity. For generating electricity from the sun we went from photoelectric cells to solar panels.

On Entertainment:

For enjoying music, we went from the woodwind pipes to alto saxophones, to trombones. From the crowd to the harp, and from the xylophone to the piano, and from the violin to the electric guitar.

On Sports:

In games and sports, we use different balls. At first they were made of leather or from a tough resin called gutta percha. Now we use inflated round balls, inflated elliptical balls, solid composition billiard balls invented by John Wesley Hyatt in 1865* replacing ivory, light bouncing balls, field tennis balls, hard dimpled golf balls…etc. Together with them, we use bats as in cricket and baseball, clubs as in golf, rackets as in tether-ball, tennis, and table tennis. Or we use long sticks as in field hockey and ice hockey, or sticks ending with a mesh pouch for catching or throwing the ball as in lacrosse. Yet the games cannot be played unless the players have suitable head-gear, leg protectants, and gloves. And none of the above ever escaped metamorphosis.

For clearing ice surfaces, we went from sweeping with the traditional brooms to the "Zamboni." It was invented by Frank Zamboni in 1944 and patented under the name "Ice Resurfacing

---

* *Thank you, Mr. Wesley Hyatt, for sparing the elephants from a destructive human fetish. Hopefully this will be extended to the rhinos, the whales, etc. We want inventors like you to follow in your footsteps. Our advanced chemistry can synthesize any substance we seek from our endangered species without sacrificing them to our atrocious ends.*

Machine". Earlier, an ice-making machine was invented by Thaddeus Lowe in 1865.

On Contraceptives:

On the male side, we went from coitus interruptus to sheep bladder condoms, to latex rubber condoms, to pre-lubricated with spermicid condoms, to condoms with additional designs to increase sexual pleasure. On the female side, fearing adultery, men obligated their wives to put on chastity belts. Now women went from intrauterine devices such as the diaphragms and cervical caps to spermicides, and from birth control bills to the morning after bill.

On Transportation Conveyances:

In one of the earliest inventions, we went from the solid wheel to the spoked wheel to the modern specialized wheels to match our multipurpose conveyances. In transportation, we went from the hansoms to cars to buses, to steam-engine trains to electric trains to levitation trains. The latter runs on a magnetic field to levitate itself above a track, thus freeing itself from friction that reduces its speed.

It may be worthwhile to point out that the success of the modern car and other comparable means of conveyances, owe their popularity to three men endowed with different innate skills. Charles Goodyear, who made possible the commercial use of rubber by his discovery of the process of vulcanization in 1839; Edward Michellin, who fitted car-wheels in 1890 with a solid rubber boundary. And to John Dunlop, who invented the pneumatic tires in 1921.

On Sea-Going:

Robert Fulton, in 1803, succeeded in propelling a boat by steam power. In 1807, he perfected his paddle steamer *"Clermont"*. A corollary paraphernalia of inventing steam-ships was the invention of ships' "iron-chain" patented by Captain S. Brown in 1808.

Divers in earlier-times used diving bells to stay underwater for a while to collect sponge. The first proper submarine took to water in 1776 during the American War of Independence. It was a wooden vessel called the *Turtle*. Its propeller was manned by the muscle-power of its sailors. It went into action against the British warships, unfortunately it was sunk. Now submarines are as big as ships. The atomic submarine is capable of launching rockets with atomic warheads from underwater. On the other hand there are submersibles that can stand great depth to carry out delicate tasks. For sailing the seas or darting over inner lakes, we went from rowing boats to galleys to galleons to schooners to steamships to hydrofoils to hovercrafts.

On Taking to the Air:

From the epochs of Icarus, man yearned to share the skies with the birds and to ascend in air. Large kites were used to look over enemy territory, as reported from China by Marco Polo, who also brought with him recipes of fire-works. Now we have hot air balloons and airships floated by helium and propelled by fans. Traveling the -"friendly skies"- changed from light aircrafts thrusted by propellers to airliners propelled by jet engines to supersonic air-liners powered by turbojet engines to swept-wing aircrafts to jump jets to space shuttles that can take astronauts to where no man has gone before.

On Armament:

To subdue the enemy, we went from slings to arrows to onagers and catapults, to guns to missiles, and from the sword to the musket to the rifle, and from Gatling guns to Bren machines of W.W.II, and from pistols to revolvers invented by Samuel Colt of U.S.A. Also from lumbering land ships of W.W.I to modern tanks, and from spitfires to gun ships to bombers. On the sea, we went from the first iron warships, The *Virginia* and The *Monitor* belonging respectively to the Confederates and to the Union Forces, to cruisers to battleships to aircraft carriers.

On Construction Gadgetries:

For hoisting heavy loads, we went from levers to tackles to tower cranes to mobile cranes. Likewise, instead of walking upstairs, we use escalators or elevators. For making holes, we went from drills to construction augers to the huge mechanical moles used in making tunnels.

We use cement mixers mounted on vehicles to build homes and to build skyscrapers. And for demolishing them, we went from dismantling them pit by pit, to the wrecking ball, to instantaneous imploding.

For roads, we went from dirt roads to cobbled roads to paved roads. For building bridges, we went from the simple truss bridges to the steel arch bridges to the suspension bridges.

On Detecting Technologies:

The methods mentioned here provide us with additive sense gadgetries to feel what we cannot feel by our hands, to see what we cannot see by our eyes, and to hear what we can not hear by our ears.

We use seismographs to detect vibrations in the earth's crust to locate the origin of earthquakes.

We use Sonars to detect sunken objects, to find schools of fish or to measure the depth at a particular place. Sonars use sound waves because they travel well in water than other kinds of waves.

Radar, which emits radio waves and processes their reflections, was used in W.W.II for detecting oncoming enemy aircrafts to foil their extraneous digression.

We also use radio telescopes to detect radio waves coming from very far away galaxies and other objects floating in deep space.

On Developing Synthetic Materials

This is a terrain to be left exclusively for chemistry books to deal with, to show how myriads of different substances came into our lives. And to show the extra-ordinary rapidity with which these products have changed to useful products that we cannot do without. Such materials include, but not to exhaustion, the following:

Adhesives, Alloys, Cosmetics, artificial Diamonds, Detergents, Dyes, artificial Leather, Lubricants, Paints, Plasters, Plastics (whether thermoplastic or thermosetting), Rubber (whether natural or modified by vulcanization), Solvents, artificial Textiles, Waxes...etc.

On Optical Achievements:

Concerning the use of lenses, we developed glasses for the near-sighted and the far-sighted, and plastic lens implants to replace human lenses lost to cataracts.

We went from the simple microscope to the compound microscope to see the tiniest of the tiny. For higher magnifications, the electron microscope was invented. It uses a beam of moving electrons instead of light and uses tiers of electric coils acting as lenses to produce the highly enlarged image on a fluorescent screen.

We went from the refracting telescope to the astronomical reflecting telescope to see asteroids, planets, and galaxies millions of light years away.

We invented the periscope, which enables an observer in the belly of a u-boat to see otherwise an obstructed field of view.

For examining the inside of a human body, doctors use the endoscopes to see inside the body without cutting it open. An endoscope has at least two fiber optic cables, one to transmit the light to illuminate the interior of the examined organ and the other to send the picture to an eye-piece, where it is viewed.

On The Art of Photography:

The first real camera was invented by Louis Daguerre at the beginning of the 19 century. It needed a very long exposure time. Then came the single lens box camera with a shutter speed of a few seconds, invented by George Eastman in 1888. Today we have cameras with lots of different parts to help us take photographs in many kinds of light conditions.

We went from black and white to color photography. And we went from still photography to Edison's Kinetoscope of 1894 to movie films taken by motion picture cameras invented by Auguste and Louis Lumiere in 1888 and played by movie projectors.

Using laser, we went from a two dimensional picture to a holographic picture with a three dimensional projection. Nowadays we have disposable single-use cameras, camcorders, and digital cameras. This latter digital imaging technology is ushering a whole new era of wireless imaging by letting the user post photos to the Web, then send them via e-mail to his friends.

On Telecommunication Inventions:

The first electric telephone was made by Alexander Graham Bell in 1876. Now we have telephones that store numbers for automatic dialing, with answering machines and with caller ID to display the name and number of callers. On the stage of telecommunication is the widely used cellular telephone that uses radio signals to send and receive calls. Following suit, the computers are joining the trend in sending and receiving information using radio signals instead of electric signals through wires.

On Audiovisual Recording:

Sound recording has its fascinating history. We went from playing voices and music on discs rotating on a turntable of a gramophone to tape cassettes played by tape recorders. With no time wasted, the Video Home System (VHS) like its predecessor the TV, became a household commodity.

Playing the videocassette, the video cassette recorder (VCR) extracts the picture signal from the picture track and the sound signal from the bottom audio track. Receiving the synchronized signal, the TV projects the image on its screen. And with no respite to the consumer, the VCR is getting replaced by the DVR, the digital video recorder, using hard drives that digitally record shows and movies.

Wagdy A. Assawah

On Capital Punishment:

We used to execute criminals by impaling, by burning at the stake, by hanging from the gallows, or by beheading by means of an axe. Latter, other devices came into use, such as beheading by a guillotine, death by an electric chair, killing by poisonous vapors in a gas chamber, or impregnating with poisonous chemicals. As to the military they carry out the sentence of death by a firing squad. However, the capital punishment is now abolished in several nations that availed themselves of civility.

A DEDUCTIVE RECOMMENDATION

To recapitulate, the aforementioned, **"Correlated Invention Conglomerates"** were samplers to confer a casual glance to evince the changes and metamorphoses in each conglomerate from the primordial to the most sophisticated. Detailing was beyond the scope of this book. Thus, the summation avoided purposely the step-wise changes taking place, e.g. in "The Telecommunications Conglomerate" and in "The Photography Conglomerate" into diverging families and lineages. It also avoided to detail the consecutive evolutionary changes in "The Sea Going Vessels," in "The Machines that Took to the Air" or any other mentioned conglomerate. This is because the extended treatment for a conglomerate into families and a family into lineages should be left to the professionals. Undoubtedly, the information available to the specialized in the art, requires the capacity of volumes to accommodate such exhaustive knowledge.

In fact, this concise narration was a calculated sprint for tempting the marathoners to expatiate. So one day, our younger generations will be prized by unique exhaustive specialized volumes on **"Inventors and Their Inventions,"** on **"Corollary Gadgetries"** following a primal invention to enhance its proffer, on **"Inventions that Went Out of Use,"** such as **The Hammod's Typewriter**, the **Spinning Jenny**, and the **Abacus** used for doing calculations; the **Aeolipile** invented in the second century, often called the first steam-engine; the **Irish Mail**, a three- or four-wheeled toy activated by a hand lever for the enjoyment of younger kids; the **Dandy Horse**, an early two-wheeled velocipede propelled by pushing with feet against the ground; the **Blunderbuss**, a short gun intended for shooting at close range without aim; the **Lewisson**, an iron dove-tailed tenon that can be fitted into a dove-tail mortise and is used for hoisting large stones; the **Astrolabe,** used to calculate the positions of celestial bodies; the **Orrery,** which shows the relative positions of our planets in the solar system by balls moved by a clockwork; the **Onager** that was used in medieval times for catapulting stones at the enemy and his enforcements; The **Franklin stove,** which resembled an open fireplace, but was designed to be set out in a room; the **Steel-Yard,** in which an object to be weighed was suspended from the shorter arm of a lever and the weight determined by moving a counterpoise along a graduated scale on the longer arm until equilibrium is attained; the **Universal Joint,** that is capable of transmitting rotation from one shaft to another not collinear with it; the **Archimedes screw,** used to raise water, which is employed for that purpose up till now in several places in our contemporary world; The **Pathyscaphe***, a medical instrument used in earlier medical practices to extricate a swallowed object resting on the fundus of a patient's stomach. Hence, for going deeper into the human body, it is made of two long metal pieces adjoined by a centralized hinge. Morphologically, the device is differentiated into three parts: a handle to work it, a long middle part to reach the stomach via the larynx and the esophagus, and an end-part forming tongs for grasping and removing the swallowed object hurting the patient.

Also needed a writ large, on the inventions of monstrous machines. The worthy candidates include: the stupendous moles used for tunneling, the gigantic loaders, the huge dump trucks, the gargantuan draglines that literally walk on footpads, the colossal

---

* *The above mentioned pathyscaphe shouldn't be confused with the bathyscaphe. The latter is a submersible for deep-sea exploration.*

stripping shovels, the mammoth bucket-wheel excavators, and the one-of-a-kind the behemothic crawler designed and designated to transport the space shuttles to where the launching pads are located.

Those specialized books will undoubtedly constitute an excellent foundation for the needed curricula meant to be taught at the long awaited for "**The Inventions Academy**". It may also be that this hoped-for Academy will, besides teaching inventiveness and researching into its various endeavors, carry the burden of bringing to life this one-of- a-kind, "**The Encyclopedia Inventiliensis**."

Capable of this humongous task i.e., of collecting and collating this enterprise, is the team at "Dorling Kindersley". For that, I allow my humble self to call upon them to adopt this one-of-a-kind project.

Wagdy A. Assawah

THE INFORMATION REVOLUTION

Many people believe that in the later years of the twentieth century, with the establishment of pervasive, sophisticated computers, we are entering a third great technological age. This transformation from the "Industrial Revolution" to the computer and information age is given various names, such as "The Third Wave" or "The Third Technological Revolution." Computer invention, which started by 1946 "ENIAC," the first computer, was a revolutionary development in the history of technology. Computers, together with their contemporaneous calculators, differ from all other machines because they have a memory which stores instructions and information. However, the memory of computers in particular: can be given different instructions, called a computer program, to do different tasks. So a computer can be a word processor, a game player, a paint box, or a musical instrument.

Trying to visualize the future is always a tricky venture. That is because nobody's crystal ball is that good. And also because if you are right on the mark, critics will dismiss your efforts as mere luck, and if your predictions are off the mark, you will be lambasted. So you will be damned if you do and damned if you don't. Also visualizing betterment in the function of a contrivance, or miniaturizing it into a friendly shape, or endowing it with more efficiency from the archetype, are all unacceptable predictions, because these are all natural eventualities stemming from our urge for improving things to make a second, a third, or a later generation.

Another factor adding to the folly of prognosticating the future of "The Information Revolution" is the presence of multitudes of inventors creating inventions that we don't know of or know about their functions. At the time of writing this manuscript, the news media were abuzz over a mysterious invention nicknamed "IT," standing for "Individual Transporter," created by the renowned inventor Dean Kamen. Rumors about "IT", which is sometimes called "Ginger," say that it rivals the personal computer and the World Wide Web in scope and importance.

The Undecided Fate of "The Information Revolution"

From pondering our contemporaneous times, one can unveil the key players shaping our present and most probably our near future too. These players are:

i.   The ever-advancing frontiers in our scientific acumen.

ii.   The unrelenting improvements in computers and other sophisticated technologies.

iii.   The exponentially growing numbers of prolific inventors in all nations.

iv.   The mushrooming electronic communication systems reducing our world to a global village.

Taking into consideration that we are living in a world in which the minds are not all benign, the exacted international laws are neither committing nor enforced, and that these laws can be easily breached by a Veto from those who claim to be the epitome of democracy is hypocrisy personified.

Also, fighting over territories marked either by a figment or parable shows that we are very far from being the truly civilized people we think we are. These situations, unfortunately, leave the prognosticators with no choice other than to be pessimistic. Surprisingly, the similar factors which caused the surge of inventiveness in the "Industrial Revolution" and which were praised for its boom, could be cursed for an unexpected doom in the already mushrooming "Information Revolution."

Consider the following, or as they say in the movie, "analyze this": Unlike the secretive, highly-elaborate fund guzzlers of the second half of the last century, i.e. The Atomic Bomb production, "Star War" technologies, the improvements in the complex war machines, building a space station...etc., there are inexpensive, simple technologies that can be used by irresponsible factions to wreak havoc.

There are now labs furnished with minimal inexpensive equipments such as autoclaves, incubators, fermenters, petri dishes, test tubes, and culture media that can produce billions of poison-producing microbes and billions of pathogenic microorganisms to generate deadly epidemics in man and livestock, or wipe fields of their harvests. And the anthrax scare of late 2001 is still alive and remembered. Other labs equipped principally with microscopes, micromanipulators, ultra-fine tubes, and needles that can artificially inseminate or manipulate cropped cells to produce clones or zombies, are another lurking disaster. Other evils also abound and they are a click away to show the anarchists and the inquisitive how to use poisonous gases, how to make pipe bombs, and other malefic gadgetries.

On account of the above, let me disavow the folly of predicting what is expected from "The Information Revolution." After all, I am not the visionary H. G. Wells who predicted among other things the TV, the atomic bomb, and the computer. Thus, he was properly called "The man who invented the future". Instead, let me concede to what is hankered from this revolution, maybe we can avoid adversities in the offings. Definitely this will be a huge task for a group of specialists and definitely it will be bordering on the impossible for one man to execute. Yet I am still willing to try the venture unswayed neither by condemnation nor by commendation. This is because doing nothing will be like fleeing from the battlefield, while providing a kernel could encourage others to participate in what can be hoped in the diversified pursuits for the benefit of humanity.

HOPES FOR HEALTH ISSUES

On Combating Genetical Disorders:

Using DNA recombinants to fight genetic disorders before they turn into debilitating diseases. The most notorious of the nearly 3,000 genetic disorders are diabetes, glaucoma, Alzheimer's disease, multiple sclerosis, Parkinson's disease, muscular dystrophy, sickle cell anemia, hemophilia, Tay-Sacks disease, dispositions to all kinds of cancer…etc. So, my misfortuned friends, please hang in there; stem-cell research has seen the light and sooner than later will be coming out of this Punic Tunnel. However, for the time being, Parkinson's patients can use the brain "Pacemaker" to control their involuntary tremors.

On Conveying Brain Stimuli:

Solving the problems of paralysis by allowing the dispatched messages from the brain to reach the muscles by jumping severed nerves destroyed in accidents. Apart from using stem cells from fetuses, there may be another venue to be exploited. And that is by

47

helping the chemical messages exchanged from one nerve ending to another by letting nanobots to do such reciprocal interchanges. Easier said than done, may be. Yet we have first to dream it and then find some way to do it. One day, walking on the moon was an unattainable absurdity, yet it happened, proving that there is no limitation to man's ability.

On Preventing Cohesion of Fetuses:

Monitoring twin fetuses in their earliest stages of formation by means of fetoscopy can end a possible cohesion. Once discovered, the common adjoining attachment is either dissolved by deep laser irradiation *in utero,* or by an operation *in vitro* to procure two separate babies. Done in the early stages of development, the tissues are bound to regain their holistic status because the cells will be in the early stages of formation and prior to the stages of differentiation into specific tissues. And this will be an operation in time that saves lives and saves agonies and expenses.

On Making Better Prosthetics:

Invasive surgeries allowed us to extract damaged parts in our bodies and to replace them by prosthetics. The technology that made the artificial heart is asked to develop durable materials to replace hip-joints, knee-joints, and prolapsed vertebral discs. To improve on the current artificial heart valves, pacemakers, hearing aids, eye implants, i.e., lenses replacing aphakias, artificial crowns and dentures, artificial limbs, etc.

Also, with our ability to transplant real organs from one person to another, there is an insistent need to develop medications to stop completely the rejection of transplanted organs in patients.

On Developing Promising Blood Substitutes:

In real blood, which in many cases may be in short supply, hosts of risk factors may spread through the bloodstream. Thus, blood can carry Aids virus, West Nile Virus, Various types of hepatitis, bacterial meningitis, and malaria plasmodia, to mention a few. Concerned over mad cow disease, the blood banks reject

49

Wagdy A. Assawah

donations from anyone who spent three months in the U.K. between 1980 to 1996. Meanwhile, synthetic blood substitutes, for which we are waiting with great exception, like Hemopure, Poly Heme, and Hemassist based on human hemoglobin and Oxygen based on a chemical called perflubron, don't carry these risk factors. Besides this valuable advantage, they are also much, much smaller in size than the red blood corpuscles. Being tiny suggests they should be able to slide past a clot or a sickle-cell jam-up to oxygenate tissues downstream.

On Fighting Obesity:

Hopes in making artificially flavored foods that taste like meat, fish, poultry…etc., using minimal caloric matrices as bulky components. Such matrices that add bulk without adding calories, could be secured essentially from enzymetically destarched multicolored leguminous seeds. White beans could be used to imitate poultry, pink beans to imitate fish, and darker beans to imitate meat. Such foods made up from cellulosic remnants or amorphous lignin, which humans cannot digest, should then be enriched by the needed vitamins and dietary supplements. Also as a prerequisite, they should be pleasing to the eye and gratifying to the palates. As such, these foods will be the awaited panacea for the obese. They will help them to lose weight, that is because the calories going in are much less than the calories spent out. With a filled stomach—being deceived by the bulky meal—will send messages to the sites of hunger in the brain, declaring a satisfied appetite. Meanwhile, the achy bodies and the breaky bones of the skeletal system will be gradually winning the battle of the bulge.

In fact, I wonder why such foods with inert (non-caloric content) matrices were overlooked by the food industry. The percentages of caloric ingredients to non-caloric ingredients in such foods could be made with different proportions, say 20% caloric to 80% non-caloric; or 40% caloric to 60% non-caloric, to meet the needs of different consumers at the different stages of their dietary regimen. Needless-to-say, exercise should be maintained, and for breaking the monotony, meals should be supplemented by reasonable amounts of fresh vegetables, fresh fruits, and nuts.

Appropriating a way for the obese to get rid of their extra weight, will always be a welcomed enactment. Hence, let me call upon our Food Industry to manufacture these non-caloric foods. And to provide our markets with this weapon designed to relief the portly from their portliness.

HOPES FOR TECHNOLOGICAL ISSUES

Concerning these issues, I am not going to recommend stepping up the production of the environmentally friendly car. Nor am I going to ask for combining the electronic gadgetries in our homes into a harmonious space-conscious singularity. This is because the titillate to industry in basic technologies and the like will keep the interest in such endeavors alive and kicking.

On Improving Air Traffic:

Using satellite technologies to regulate air traffic can prevent collisions or other mishaps in the vast blue yonder, or on runways. Had it been used, it could have reported the debris on the tarmac that caused the latest concorde tragedy.

On Insuring Safety On The Road:

This can be done by providing cars and all sorts of moving vehicles with sensors to keep them apart by a safety margin and to keep reminding the driver to stop tail-gating the other car in front of him.

On Developing Benign Atomic Energy:

The atomic energy is now available, yet it is not widely used, because it needs to be safer and more benign. The Chernobyl disaster in Ukraine is still a painful experience that cannot be overlooked. So it is a priority to develop safer atomic plants to protect populations and environments from the harmful radiations.

On Bringing Sun Rays into Our Homes:

We can bring sun rays into our rooms by providing our homes with ducts accommodating fiberglass tubings that collect sun rays from tops of buildings and send them to our inner rooms. Such

Wagdy A. Assawah

technologies are used in Japan and the U.K. They need only to be standardized practices in our U.S.A.

On Stopping the Use Of Radio Active Explosives:

We should develop non-radioactive explosives, so the leftovers cannot go on killing the innocent we went to protect from the criminals of ethnic cleansing. It is ironic to leave behind cancer-producing remnants that kill slowly those we want to rescue from a swift death, or those liberated from a tyrant.

On Destroying Oncoming Enemy Missiles:

Hitting an oncoming enemy's missile with an outgoing missile is comparable in difficulty to hitting a bullet with a bullet. This makes the practicality of its destruction a dubitancy. Should we then hope for another alternative method? Using concentrated laser beams for the purpose may be a plausible alternative. For preference, they can travel with speeds comparable to the speed of light, can get focused on an oncoming projectile, and with their penetrative power, will be able to destroy the enemy missile. Also let us not forget, that

54

the enemy will not send one, he will be sending many. For that, we should have as many batteries to put up with those unwelcomed guests and to keep putting them in the crosshairs of the laser guns to destroy them before reaching our borders.

HOPES FOR ENVIRONMENTAL ISSUES

To understand how we can harmoniously live in our global environments and in the different ecosystems, it is plausible to explain our stance in this world of ours.

Though the universe may look chaotic, yet there are recognized patterns regulating its different components. Studies showed that what pulls the strings of macrocosms or microcosms; the vivacious existence of living beings or the defunct existence of nonbeings, is a group of orchestrated rules. Hence, planets and apples obey the same forces of gravity. In physics, elements function according to delineated properties. In chemistry, molecules react according to poised valencies. In biology, living beings perform according to predetermined doctrines. However, the previously mentioned rules, though mindless, yet they are what keep the atoms and galaxies in sync, and what keep the animated beings performing according to pre-ordained perimeters. Likewise, the unwelcomed natural disasters from volcanoes spouting lava, to earthquakes, to floods, to forest fires, to avalanches, to tsunamis, to tornados, all happen according to natural forces governed by predetermined rules. On the other hand, man-made disasters from drought, to desertification, to global warming, to erosion of coastal areas, to overpopulating our habitats exhaustively, to hunting down to extinction our prized fauna, to plagues whether caused by microbes, protozoa, insects, or rodents; are all due to messing with Mother Nature's doctrines.

Unfortunately, we have no jurisdiction on natural calamities, and nature will not compromise its rules to spare us humans because we are all casualties of evolution. Hence, we may end up like the Dinos, i.e. **Extinct**, and the world will neither shed a tear over our demise, nor will it notice the disappearance of the ultimate swarmers who populated this speck of dust.

However, for a congruous co-existence we must strive to:

- Achieve a scale-down of our ever exponential increases in number, to equate with our territorial limits.

- Achieve healthier Global Environments.

- Establish Reconciliatory Equilibria for our Ecosystems.

- Learn to appreciate the Triumvirate of Man, Fauna, and Flora.

On Achieving A Scale-Down For Our Exponential Increases:

In spite of our brains, whether we like it or not, we unconsciously obey the same rules that govern grasshoppers and ants, rats and rabbits, bisons and moose, sharks and Orcas. The relentless waves of the poor and the oppressed, fleeing their countries to the technologically advanced rich nations are caused by both overpopulating their heartlands and by injustices among nations. And that we as nations, whether rich or poor, are victims of "**Rubbing-the-Elbows**" syndrome.*

---

* *This phenomenon is known in migratory desert locust, **Schistocerca gregaria**.*

It is no secret that we continued to bust the seams. And to find "Blue Planets" in outer space for accommodating our ever increasing numbers, is up till now, a tall order. Hence we have to solve this big issue of our contemporary world here and now. And if we don't willingly, put the brakes on replenishing the earth, the competition over the greener pasture will inevitably get us thinned the hard way, i.e., by grinding wars that could pale Armageddon.

This war ushering "The End of Times," that was expected by our ancestors and accepted by us without repining, will not be due to the second coming of "Christ" to vanquish the "AntiChrist" and his seven-headed beast(s), but it will be due to the quandary of over-crowding this circumscribed planet of ours. To the author's scrupulous mind, the freedom of raising large families, that was enjoyed by married couples from the dawn of history as an ipso facto privilege, is now coming along as a threatened luxury which humanity cannot afford to practice any more. Unfortunately, the rapidly accelerating rates in our concrescences will force us to choose between two equally afflictive solutions. It is either to continue to profuse, consequently get culled by agonizing wars, or to deliberately reduce to avoid taking the cudgels against each other and hopefully live happily ever after.

Ironically, some nations, for a solution to their congested cities and to their inabilities to cope with the needs of their ever-increasing populations, choose to send their surplus to America*, i.e., **export the problem**. Unfortunately, America's acceptance of the surpluses from other countries is a policy dictated by the moguls of industry who take and distribute bonuses in seven digit numbers. Such easily earned bonuses represent the differences between the hard earned minimum wages and what should have been paid to the toiling extra load on our ark. Obviously the motto of certain "Corporate Executives" is "To hell with the devastations befalling the country as long as our pockets get deeper and deeper."

Deplorably, the harm didn't end with this peccant outcome. Besides sitting at the top of the "Exploitation Pyramid" enjoying their voluptuous bonuses, they also orchestrate their corporate pecuniary. From their Ivory Towers, and with their thumbs on the flow of dough, they will know when the stocks will bottom out and then buy on the inexpensive, and know when the stocks will crest and then sell on the gratifying. This brings to the mind Mr. Bob Vila's pit-saw cutting on its way up and cutting on its way down, as if the bonuses weren't enough to quench their greed. This invented

---

* *And to Mr. Donahue's question, "Should the U.S. close its border?" I say, "Appropriate, but not to exclusion!"*

despotic corporate chicanery is literally transforming the bona fide investments into a high handed gambling extend, in which the only winners will be the shady characters running the bank. Or worst, a lottery in which the predetermined winners will be none other than the pietistic "Corporate Executives."

Harkening back to the deluge of surpluses coming to America, the plea seeking an answer is: To accept or not to accept? That is the question! Yet, if we decide to accept, where is America going to send its surplus? For an answer, and because of the unavailability of "Blue Planets" and the lack of means to take us there, I say, "Heaven knows, Mr. Chips!"

Up till now, our "Rubbing-the-Elbows" is manifesting itself, fortunately not by wars, but by aggressive migratory movements. These phenomenons of **"Fleeing in Swarming Congregations,"** are new alarming blights. As a testament of this adversity, the Asians locusting on cargo ships heading for Australia; the Eastern Europeans crossing the Adriatic Sea under the cover of darkness to land on Italian shores; the slowly but surely line-ups from African nations infiltrating into Spain and beyond. The ferocious night raiders of the Euro tunnel ganging up for a stow-away ride to The British Isles; the obdurate burrow makers smuggling Mexicans and drugs to U.S.A., meanwhile, the affluent and the aided by terrorist monies choose to go to Canada then to slip through our cottage-cheese northern borders to the heart land.

So are we going to convince ourselves that this is a fading fad? Or are we going to give heed to the **signs** and find a way out of this pandemic dilemma?

It is now or never, that the "Big Brothers" who turned their backs for decades and even exploited these nations for their own well-being, to act expeditiously to rectify the conditions in these unfortunate countries. In so doing, we will be defusing to a great extent these mushrooming migratory movements. Because if we do not act now, our already over crowded cities and suburbs will be confronting myriads of unforeseen problems. It is a condition in which it is wiser to challenge this global problem together than to freelance the consequences disaffiliately.

In short, prudence decrees that we should hang together, otherwise we will be hanging separately.

Wagdy A. Assawah

On Achieving Healthier Global Environments:

Our abusive activities to our global environments, i.e., the gaseous envelope, the lands and the seas can bring on us, the inhabitants of this earth, a calculated expiration. And this is a counseled lesson from an insignificant micro-organism. **A bacterium!**

Under favorable conditions, a bacterium can divide and thus provide its equivalent of a human generation about every twenty minutes. Somebody has computed that in some very short period, a single bacterium could give rise to a mass of bacteria of greater volume than the whole earth. Fortunately for us, this will never happen. And here is why.

What did then prevent the bacterial colony from reaching this phantasmagorical size? One may say that in the process of multiplication, the bacterial colony depleted its sources of nutrients and consequently, the colony died out. Though this may be a contributing factor, yet this by itself is not the principel causation of the bacterial demise. The fact is, the death of the bacteria was mainly brought about by the poisonous excreta they produce in their surroundings. This phenomenon of an organism dying in its own offals is a decretum of Mother Nature embracing all living creatures from bacteria to man.

The world population has doubled between 1950-1996. These continued exponential increases in our numbers will eventually put us on a collision course with the environmental limits. However, not only are we depleting the resources of, so to speak, our petri dish but we are continually abusing and poisoning Mother Nature's environments.

So, we are spewing all sorts of harmful gases into our atmosphere. Nowadays, the air in many cities all over the world is unbreathable, dominated by smoke and chemical fumes (i.e. smog). Acid rains are also a commonplace. The ozone layer shielding us from the harmful ultra-violet rays is losing its efficacy due to the continued emanation of chlorofluorocarbons. Heat pollution from our countless internal-combustion engines is ushering a quicker global warming that will result in higher sea levels threatening to submerge uncountable number of coastal cities.

As to our oceans and seas, we are over fishing them, devastating their floors, and dumping our sewage in them. Rivers, inner lakes, and swamps get also their share of poisonous chemicals as by-products ditched in them from our industry.

Our cultivated lands, whether in the valley or on the slopes of hills, were not luckier than the atmosphere or the oceans and seas. We are exhausting them by our cultivation, poisoning them by fungicides, insecticides and by harmful chemicals carried by irrigation waters. Now our ever-expanding concrete jungles and our billions of wider and wider asphalt roads are shrinking them down. They are also obstructing the free gaseous exchange in the soils. Thus, mother earth's stomach i.e., the smothered soils, over and above of being deprived from the oxygen needed by their microorganisms to complete the digestion of our offals, will also get faced with increasing amounts of harmful refuse beyond their abilities to decompose.

In earlier times, forests covered areas of up to 45% in the different continents. Being the lungs of mother earth, they give us the benign oxygen and assimilate the carbon dioxide that we relentlessly generate, and causes the green house effect. Now the forests are shrinking at an alarming rate to a critical insufficiency. And together with losing our timberland, we are also losing an assortment of valuable zoogenic creatures.

Unfortunately, not knowing better, the generations that came before us, believed that our gaseous envelope is illimitable and their emissions cannot impair it. Our seas are perdurable and as such they will continue for good, to be our providers for sea-food and all sorts of useful products. Our cultivated lands are invincible, so they can digest our offals and can go on giving us their bounty of grains, fruit, vegetables, fibers…etc. Our forests are imperishable because the trees are self-generators. And our cosmic earth, is infinite and can sustain our ever-increasing numbers.

The previously inherited convictions may explain why our unaware governances give no heed to the pleas of guidance begged by our Scientific Environmentalists. Their forewarnings are getting more and more important. That is because the subsidies we get from nature cannot withstand being overdrafted time and time again. And because our ever-exponential increases in number in every nook and cranny, are amplifying our deleterious effects on our global environments which, at one time we thought to be infinite; meanwhile, for an admitted fact, they are circumscribed.

The previously mentioned bird's eye view on how we are abusing our global environment shows that our mismanagement will eventually catch up with us. Thence, Mother Nature will obligate

Wagdy A. Assawah

us to pay for arrogantly ignoring the signs. No wonder then, if our demise will be comparable to that of the mindless bacterium. That is, dying by our own caustic excreta! Or shall I say by our stubborn idiocy?

On Establishing Reconciliatory Equilibria For Our Ecosystems:

Under this heading, the stress will be on highlighting the importance of equilibria in maintaining the comparatively localized ecosystems. This can be helped by the diagrammatic representation given in Figure 1, showing the intertwined relationships of the key players in those delicate systems. Hence, if one inhabitant becomes incompatible with the others, the balancing perimeters keeping this particular ecosystem in a unison will malfunction, resulting in a collapsed habitat.

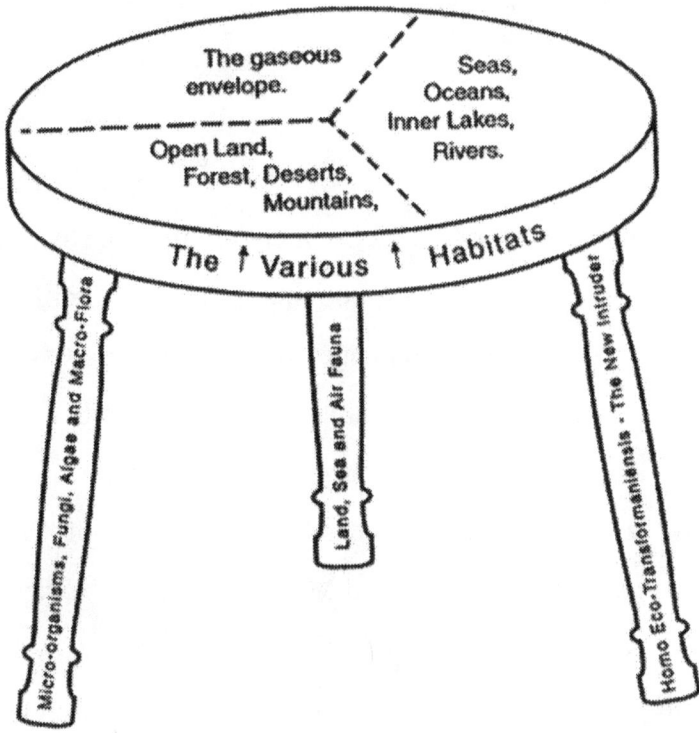

Figure 1 - Showing a three legged stool to symbolize the equilibria existing between the various inhabitants and their habitats.

In the drawing, the seat represents the different habitats. Meanwhile, the legs represent the co-habitants of the different ecosystems within each habitat. Thence, one leg emblematizes modern man's strenuous activities on the system. The second accounts for the effective participation of the phytonic beings. And

67

the third portrays the vivacious bustling of the rest of the zoogenic creatures.

It may be worthwhile to mention that the earlier homo primates who came before us, populating the globe millennia after millennia, did very little damage to the environments. In those epochs, Mother Nature was able to cope with their inevitable activities and to rectify the harms done by them. On the other hand, the nowadays illegitimate mass destructive activities of *"Homo Eco-Transformaniensis*"* are beyond Mother Nature's abilities to undo. And unless we clean up our acts, the different habitats will become unlivable for our future generations. So we should stop taking our Mother Nature's ecosystem sanctuaries for granted.

Henceforth, notifications and few humble proposals are forwarded to claim back our abused ecosystems from the brink of destruction.

---

* *Homo Eco-Transformaniensis: Assawah, is a pseudo-scientific name. It was coined to stress the malefic changes that man can inflict, on established ecosystems, to serve his extraneous ends.*

On Introducing Newcomers To An Ecosystem:

As a rule, for protecting established ecosystems, we have to carry out extensive studies before introducing a newcomer into an established ecosystem. Be it a microorganism, an insect, a plant, an animal, a shellfish…etc., because a newcomer may overrun the endemic species, upsetting the existing equilibrium working for that particular ecosystem. The examples delivered here-in are meant to explain and to justify the mentioned monition.

In a nutshell, we are all familiar with the pandemic afflictions caused by the bubonic plague of Medieval Europe, and the introduced HIV in recent times. In our U.S.A. we are also familiar with the bio-devastations caused by the Mediterranean fruit fly, the fire ants, the African Killer bee, the Hyacinth chocking waterways, the purple loosestrife strangling the wetlands, the white-tailed dear becoming a pest, to mention a few.

In Guam, the introduced tree snake annihilating the songbirds, in New Zealand, the ferrets and weasels infesting their new habitats, in Australia the Opuntia plant and the introduced rabbits exploding out of proportion. Later, the rabbits were targeted successfully with the myxonmatosis virus to cull their numbers.

On Fighting The GreenHouse Effect:

Curbing the greenhouse effect by cutting down on the use of fossil fuel and by keeping felling forest trees to a minimum while at the same time, replanting the denuded areas, and by also finding ways to reduce the harmful emissions from our cars and factories. For a disturbing fact, while our population doubled between 1950 and 1996, the number of cars increased tenfold, contributing also to heat pollution.

On Mending The Ozone Hole:

Measures to slow down the destruction of the ozone layer by prohibiting the use of chlorofluorocarbons are not good enough. For an active measure, mending can be done by deploying balloons filled with man-made ozone and emptying them in the stratosphere, where it is needed. Though the ozone hole may be as big as the continent of Australia in dimensions, yet what encourages the endeavor is the thin nature of the ozone layer and that gases are able to expand to occupy larger areas, indicating that the suggested idea could pay off, if we keep at it.

On Depleting Our Seas From Their Bounty:

Unfortunately, we are over-fishing our oceans and inner lakes, impairing their productivity. Also the multi-layered sea-floor did not escape our devastating activities. And the ways we use to catch fish, speak volumes of our destructive methodologies in depleting the seas from a good source of food.

Unappreciative of the consequences and thinking that bigger is always better, we are now applauding the commercial fishing, using the long-line fishing, the gill net fishing, and the purse-seine used to round up schools of fish. Also, we did not give respite to the ocean floor and we are damaging it by using the trawl nets to snare cod or shrimp and by using the steel-teeth dredges to scoop up scallops and oysters. The continuous abuses cause devastations to the fragile ocean floor habitat, the feeding grounds and hiding places for juvenile fish. The best policy is, "Everything in moderation."

Simply by burning the candle at both ends, we are not only hurting ourselves but also our future generations.

On Fighting Water Pollution:

Detoxifying waters coming from agricultural fields carrying pesticides and detoxifying by-products from factories carrying remnants of heavy metals before reaching our rivers or our territorial sea waters, killing all sorts of creatures from planktons to whales. This is specially important because heavy metals going into the food chain will eventually end up in our foods, destroying our organs and poisoning us.

On Reclaiming Land Lost To The Sea:

We need to find ways for protecting our shores from erosion and our harbors from getting submerged by rising sea-levels. And if we have to live with the ascending tides, then we should learn how to make good use of our continental shelves, our new unexplored inner space.

On Developing Clean Energies:

Environmentally friendly sources of energy are needed instead of relying on fossil fuel that we are depleting to the point of extinction. This could be done by perfecting our solar batteries and finding ways to store their energies procured in daytime to be used at nighttime. By developing new technologies to obtain energy from the inner hot layers of the earth's crust; By using the action of tides or the difference in sea water temperatures as possible sources of non-polluting energies. Thus, we can rely less and less on burning fossil fuel contributing to the greenhouse effect.

On Combating Forest Fires Effectively:

Protecting our forests by installing powerful cameras mounted in satellites circumventing the globe to report fires in their primordial stages before they break out into hard-to-control infernos. In so doing, we will be sparing mother earth lungs, lessening the disappearance of rare plants and protecting endangered animals from a terrible fate. This will be more valuable if other nations come together to be members in this universal project. Hence, all countries will be benefiting from the reduced expenses that come

73

Wagdy A. Assawah

with deploying man power for either combating a sweeping disaster or for replanting the denuded areas.

On Heading Off Desertification:

Staving off desertification, especially in territories juxtaposed to seashores, can be done by desalinating seawater by economical gadgetries to provide palatable water for drinking and water for maintaining a plant-cover to hold the upper cultivated layer from being eroded by blowing winds. As to sea-shore cities threatened by drought, e.g., cities in the Iberian Peninsula, Morocco, southern Australia etc., these can be helped by hauling and harnessing the iceberg's potentials. Innerland areas suffering from drought can be helped either by diverting tributaries or seeding the clouds to release their load of moisture or by digging wells to get to the underground water.

On Appreciating The Triumvirate Of Man, Fauna And Flora:

This is the most intricate of all the reconciliatory equilibria supporting the functionality of our ecosystems. Its holistic benefits encompasses every element in the system, and that is why its discussion was kept to the last.

Mother Nature has trusted us *Homo sapiens sapiens* to be the keepers of animals, plants, birds, insects, micro-life, etc. Given highly circuited brains, we were burdened with the responsibility of handing them down to the coming generations unscathed. It took Mother Nature hundreds of millions of years to make them the way they are today. And it took her eons to perfect the sensitive balanced ecosystems they live in. Such balanced systems comprise; forest ecosystems, savanna ecosystems, desert ecosystems, tundra ecosystems, riverbank ecosystems, continental shelf ecosystems, mangrove ecosystems...etc. The acclimatized creatures living in a particular ecosystem thrived, reproduced, died, got recycled, and from the end product came the new living generations after generations. All these systems with their zoogenic creatures and phytonic beings were and still are parts of valid equations that keep working for us. And as long as such systems keep delivering, our unwelcomed quick fixes to save in the short term will make us pay more in the long term.

Animals that arrived before us and lived in those ecosystems are the rightful owners of those habitats. Yet we the recent encroachers, invented the "deed" to claim the lands to be ours and ours alone. A red-Indian chief seeing herds of Bisons slaughtered to extinction decried the carnage, saying, "What is man without the beast!"

In his bleeding slogan, one can perceive that this overwhelmed chief has sensed a holistic balanced ecosystem wasted for no reason except greed. Unfortunately, this foolish wasting of animals, plants, birds…etc., together with upsetting the delicate balances of such ecosystems, is still at large. Such actions leave behind barren lands difficult to reclaim, while we are watching and with nobody raising a finger to stop the heinous acts, taking refuge in the N.I.M.B.Y. (not in my back yard) syndrome.

Was this because, we were ordered to be fruitful and to replenish the earth? Unfortunately, Gods do not speak to us anymore to say "Stop, you have busted the seams." Yet we have matured enough to come to grips with the fact that each area has a calculated capacity, and that its animals, plants, birds and insects are living in harmonious co-existence. And killing the animals to expand on their territories is wrong. And to raise crops on denuded forest areas is no plausible solution. Because in so doing, we are undermining the equilibria of such natural ecosystems and some day our future generations will face the consequences and blame us for our selfishness and our mismanagements.

If we are unable to appreciate the equilibria sustaining these ecosystems, then we should take heed of what has happened to the Polynesians of Easter Island, and to the Mayans who lived in an area extending from the Gulf of Mexico in the north, to the shores of Guatemala bordering the Pacific Ocean. The first, i.e. the Polynesians, destroyed their vegetations to the last twig, leaving behind hundreds of monolithic statues of human form. On the other hand, the Mayans lived in a lush rain forest brimming with generous vegetations rich in wild animals and game hunted for sport and food. Over several centuries, they built countless cities and raised crops on huge cleared forest areas, upsetting the equilibria of their ecosystems. Over the passage of time, felling of trees by the millions caused the shrinkage of habitats for wild animals and game. With less transpiration, the clouds were tightwad with their precipitation, impairing crop maturation and production. Seeking food, the Mayans hunted down game and wild animals to the point of extinction. Thence, they fought over land, each thinking it was greener on the other side of his metropolis. At last, times came for Mother Nature to exact its punishment. Coming to the New World, we knew about the Mayans, their existence and their demise from the standing buildings and temples that were overgrown by the forest plants that gained their vigor in the absence of their abusers.

Similar calamities caused by mistreating established ecosystems in the Old World also abound. Visitors to Ancient Egyptian monuments can see on their walls a variety of animals, birds, and arthropods. Depicted were hippos, crocodiles, lions, monkeys, jackals, snakes, gazelles, cows, cats, dogs, horses, an assortment of birds, scorpions, scarabs…etc., but not a single camel was ever sketched. This is because the wide strip of land lying between the Sahara Desert proper and the southern shores of the Mediterranean Sea wasn't a desert in need of the flattened hoof of camels. Instead it was a lush green ecosystem teeming with game and wild animals. This land with its varied plant cover was the provider of grapes from which the Romans made their wines, of olives from which they extracted their oil, and of grains from which they made their daily bread. Trees of figs, pistachio, pomegranate, and other annuals were all in abundance.

It was the low grazing animals that came with the Arabs invading North Africa in 670, that destroyed this ecosystem to no return. Goats, lambs, and camels ate the plants down to the roots. The fertile top soil formed over the millennia by Mother Nature was thus exposed and blown into nothingness. The winds which weathered away the fertile top soil were also the winds that brought the sands from the south. Decade after decade, the ever encroaching sand dunes came to dominate the landscape, leaving only the huge Roman reservoirs to lament the ending of an irreclaimable balanced ecosystem.

For a first impression, one may think that these are ecological dilemmas unrelated to inventions and inventiveness. However, this is farthest from right. Once we invented to enjoy life, now we have to invent to sustain our lives. Thus, finding ways to keep the natural ecosystems operating, and inventing ways to keep the delicate equilibria implemented is a must. Also we shouldn't let one element in a particular habitat to overrun another element in this particular habitat. Because a *"laissez faire laissez passé"* attitude is nothing less than, "We don't care as long as the deluge comes after we leave this world."

Unfortunately we heed only the logistics of fear when Mother Nature's wrath is sudden and abrupt. But we don't pay heed to the wrath of destroying habitats because of its laggardness. Unfortunately, due to our shortsightedness, this wrath is going to befall our future generations. And in my opinion, doing the crime and letting others do the time is a dastardly act.

The environmental shambles* we are now facing ask for establishing specialized "**Environmental Institutes**" to search high and low for cures ailing our environments. Making our world safer for us and safer for our forthcoming generations, besides being a noble cause, is also a pressing need. And I see no other cause that can have precedence over this meritable cause.

We lost the mammoth and a host of other animals and plants to oblivion. We want to lose no more on our watch. We want to leave the animals, plants, birds, and other beings in their balanced ecosystems to the coming generations so they can find in them a bounty of useful usages. It will be betrayal to the unwritten commitment to let down our magnificent fauna from the elephants

---

* *Certainly integrating our sine qua non Greenpeace in the contemporary federal system will be an appreciated act. That is because G.P. will be there to provide the correctional concepts favoring balanced views on environmental issues at the congregational level.*

to the earthworms the first plowers of soils and from the whales to the mother-of-pearls or to let our beautiful flora from the giant Sequoias to the smallest grass to go the way the Dodo went. We want our future generations to see the whole gamut not in a book, or in an animation, but in nothing less than the real flesh.

Thus, Wise I Rest My Case...!

Visualizing that we can destroy our world by atomic bombs and weapons of mass destruction, H.G. Wells inscribed on his tombstone, "Damn you all, I told you so." Hence continuing to denude our forests, to hunt down to extinction our precious animal companions, to poison our lands and seas and to pollute our gaseous envelope, warrants me to inscribe on my tombstone, "Damn you all, I told you not to temper with the equilibria of our established ecosystems."

CHAPTER 2

INVENTIVENESS AND CLASSIFICATIONS

THE DIFFERENT INVENTIVE BEHAVIORS:

There is no school up till now to teach inventiveness, yet there is an inventor in each and every one of us. Surprisingly this disposition, i.e. the inventive behavior, doesn't exclude other creatures. In fact, different animals, according to their evolutionary status, display progressive degrees of inventiveness. Man across the ages has seen such dispositions in birds, rodents, other animals, and primates.

The Passive Inventive Behavior:

Sea-gulls with shells in their beaks, fly high, then drop their catch on hard rocks to bust the shells open to get to the tender meat inside. This disposition is considered to be the most primitive expression of inventiveness. Scientists of Animal Behavior have noticed and studied similar acts in birds, rodents and the like, either to get food or to escape eminent dangers such as electric shocks.

The Active Inventive Behavior:

In this disposition, the concerned animal uses a simple instrument available in his surroundings to achieve a desired outcome. So monkeys, gorillas and other primates use sticks to move branches to get objects that are out of their reach, or use flexible strips of bark to fish for termites to eat them. Chimps were also shown by primatologists to carefully select flat stones to use as anvils on which to crack hard nuts with pieces of wood to get to the edible kernels to enjoy.

Wagdy A. Assawah
The Deliberate Inventive Behavior:

Privileged by a highly circuited brain, man is the only creature who can assemble an assortment of components to make a complex gadget to do a certain job to satisfy a particular need.

A would-be inventor has to be, motivated by a necessity, exhibit premonition for a resolve, and possessing interdisciplinary knowledge to help him execute a consummated novelty. Yet inventiveness is a God, given gift. Those who have it, display what one may call the **"Innate Deliberate Inventiveness"**. And those who don't have it could be taught to acquire it, hence the appellation, **"Acquired Deliberate Inventiveness"**. In fact, the suggested **"Classification of Inventions"** forwarded herein reveals, that it has rewarding possibilities by showing that inventiveness could be learned. And that through systematic thinking, one can train individuals to follow certain pathways to come up with desired useful inventions.

CLASSIFICATIONS AS TOOLS FOR PERFECTING OUR KNOWLEDGE

Looking at things around us, we try to find similarities or dissimilarities to help assigning them into different categories. This is done to facilitate studying the subjects under consideration. For a fact, both the inanimate and the animate worlds did not escape this obsession of ours, i.e. categorizing and classifying.

As to the **"Inanimate World"** it comprises the different elements found in nature. According to their atomic weight, they were classified into **The Periodic Table of Elements**. Finding unoccupied cubbyholes in the table suggested to the Physicists and Chemists the occurrence of elements that were neither known to man nor discovered by him up to this point in time. This fact enticed the scientists to look for these "Missing Elements." Surprisingly, through spectroscopy of sun rays, they knew of their occurrence in the universe before they were proven to exist on earth. And this was an undoubted reward that was profited from the accessibility of the "Table."

As to the "**Animate World**", the starting point for the biologists begins with the presence of macromolecules in the early Precambrian seas. These macromolecules, whether nucleotides or gene fragments, presumably came to earth on meteorites from outer space, inseminating the early seas contemporary to the splitting of the super continent called **Pangaea**. Consequently, this notion pushes the question of creation to another faraway planet. It also insinuates that other worlds in our **Milky Way**, or in other galaxies, may have been inseminated by such meteorites, substantiating the idea that we may not be alone in this universe.

However, some biologists attribute the presence of macromolecules in the early Precambrian seas to lightening discharges in the Precambrian atmosphere rich in nitrogenous gaseous compounds, catalyzing the formation of such macromolecules in abundance. To the biologists, such happenstance, i.e. the presence of macromolecules by either heavens marrying the earth or by immaculate conception brought about the electric discharges is as valuable to them as the "**Big Bang**" theory is valuable to Astronomers.

From these Precambrian seas, or soup rich in macromolecules and nucleotides, ensued uncountable creatures and plants of all sorts, sizes and shapes. And by uncountable evolutionary processes spanning over 4600 millions of years emerged the "**Zoogenic creatures**" climaxed with *Homo sapiens sapiens*. Also emerging were the "**Phytonic beings,**" culminating with a multitude of flowering plants. It was then the job of our biologists to classify the assortments of zoogenic creatures and phytonic beings on hand. Hence, the animate world was broadly divided into the Animal Kingdom, the Plant Kingdom, an "Intermediary Groups," comprising mostly of microscopic and submicroscopic organisms. These kingdoms and the intermediary groups, in their turn, did not escape the splitting into Classes, Phyla, Orders, Families, Genera, Species, and Sub-species.

The tendency to assign the elements of the inanimate world in cubbyholes as in "The Periodic Table of Elements" and to split the animate beings into groups emphasizing the evolutionary trends ended up with an unexpected yet exciting discovery. This discovery suggests that the boundary between the "**lifeless matter**" and the "**life-full beings**" indicated an uninterrupted continuity between these two apparently contradictory components. It was the tiniest entities i.e. the viruses, that bridged this discontinuity occupying, so to speak, this "no man's land." They (the viruses) display the ability to crystallize, sharing this feature with metallic salts, a characteristic of the inanimate world. They also display the ability to multiply in invaded host cells, a characteristic of the animate world. This amphoteric nature coadunated the inanimate and animate worlds, proving the unity of creation in an uninterrupted continuity. Hence, the possible existence of an intrinsic **Mega-Cosmic Formula** equaling in its intrigue to the imaginary "**Philosophers Stone**" sought after by our earlier alchemists.

CLASSIFICATION OF INVENTIONS

Barely enough was mentioned about the classifications of both the inanimate and the animate worlds. What, then, about the inventions, the subject of this book? And can the suggested systematization, offer new horizons to the inventiveness endeavor? And can it corroborate the author's claim that inventivity can be learned!

People involved into classifying things are themselves susceptible to classification. They fall broadly into two main groups: **"The Lumpers"** and **"The Splitters"**. On one hand, the lumpers, as the name implies, tend to make fewer groups containing numerous subordinates in each group. Meanwhile the splitters tend to sort out their studied subjects into numerous groups containing less subordinates in each group. As far as I am aware, the system used by the Patent and Trademark Office "P.T.O" has many art groups. These in their turn are divided into divisions and the divisions into subdivisions. This indicates that the P.T.O. system is a splitter's system and not a lumper's system. Yet this shouldn't prevent others from suggesting other systems that may route the efforts for would be inventors to unexplored territories in which they can try their creativity.

Examining many of the already existing inventions and many of the recently approved ones by the P.T.O., the author recognizes four clearly distinctive classes of inventions, namely:

1.   The Transfiguratory Class of Inventions

2.   The Amendatory Class of Inventions

3.   The Assembleatory Class of Inventions

4.   The Conceptual Class of Inventions

The Transfiguratory Inventions

This group comprises the simplest inventions. They are usually initiated by established industries, and usually represent continuations of earlier innovations. Yet they stress metamorphic changes in the whole structure of an already existing product to keep the interest of consumers unfaltering. The introduced change claims either a more efficient job or a more suitability in handling of the new generation than its predecessor. Notable examples of this category comprise the roller blades and their transfiguratory, the skating board; The motorcycle and its miniature metamorphosed, the motorized scooter; The tricycle and its related bicycle and the metamorphosis of the latter into the tandem; The glider and its metamorphosed hang glider; The autogiro and its metamorphosed helicopter. Examples also abound in cookware, the various tools used in industry like hammers, spanners, screwdrivers…etc.

Other products used in our daily life that were targeted by transfiguration include the shaver. About 70 years ago, the disposable shaver came into use to replace the thin, double edged blade, invented by King Gillette in 1903. The first disposable blade had one efficient blade credited for a job well done. Yet to arouse the interest of the buyers, the industry came up with a metamorphosed shaver having two blades, claiming that one blade holds the hair while the other cuts it. And now the market is flooded with a third generation under the name "Mach 3" with a mysterious third blade. The fact is when you have more than one blade, the pits and pieces from the shavings will get jammed between the superimposed blades, affecting the efficacy of the shaver.

Also, our daily toothbrush did not escape this transfiguration. One may mention the "Oral-B Criss-Cross tooth brush" claimed to have two types of hair tassels crisscrossing each other. So one type is claimed to clean while the other is claimed to sweep. Another example is the "Hinged Tooth Brush," claimed to bend in conformity with the contours of the human jaw. A third is the "Navigator tooth brush" with additional flanking rubbery appendages claimed to massage the gums while you are brushing your teeth. Last but not least came the motorized tooth brushes exemplified by the battery-operated "New Colgate Acti-brush," claimed to have three-dimensional action, and the battery operated "Crest Spin brush."

The Amendatory Inventions

This group addresses making changes in subservient parts of an established invention. Thus, they will be concerned with improving an integral component in an already existing machine for improving its efficiency, its speed, or to make it environmentally friendly...etc. No earlier invention has escaped this reform. That is because humans enjoy since the dawn of history an innate urge to excel over themselves. The products of this driving zeal, is usually called the second generation, the third generation, and so on. Targeted by such amendments, one may mention trains, cars, airplanes, communication gadgetries, guns, ships, computers…etc.

In cars, one may improve the rack and pinion for better steering, the differential to go smoother around corners, the fuel injection system for increasing gas mileage ratio and to be gas sippers instead of being gas guzzlers, the efficiency of the gearbox to achieve slick cruising. And such as changing the wing structure in airplanes to increase its lift or changing its shape to reduce drag to enable high speed flight or providing durable cables to move the flaps on the wing and the rudder to steer the plane more safely, or to replace the jet engines with turbofans for superiority in power and efficiency in fuel economy. And such as improving the resolution in optical instruments, whether be it a microscope, a telescope, a binocular, a camera…etc. Also improving a microchip in a mouse or in a computer…etc. Such amendatory inventions are part of our daily life. We cannot do without them and the innate faculty to perfect them will keep us improving these inventions to produce more and more sophisticated generations of these useful inventions.

The two earlier discussed types of inventions, "**The Transfiguratory**" and "**The Amendatory**" inventions, constitute gadgetries that had gained proven acceptance from the consumers.

Though reiterative, they form the backbone of established industries, and the changes in them are always welcomed by the manufacturers. The industry has antecedently the needed assembly lines, and amendments for a better product won't need much changes in their machinery or assembly lines. For another reason, the new products will be awaited for, insuring a lucrative market with rewarding revenues.

The Assembleatory Inventions

This group comprise inventions in which two or more different components are integrated to form a new assembled product. Examples of these are numerous and proved to be successful in our daily life.

The first motorcycle was built in 1885 by a German, Gottlieb Daimler. He fitted one of his gasoline engines to a wooden bicycle frame. Today motorcycles are more complicated machines. Adding a sail to a surfboard, the sailboard came into existence. Adding two inflated pontoons to a bike, brought about a shuttle-bike to ride across mountain lakes. Adding airfoils to a floating boat made from it a hydrofoil. Adding a contouring skirt to a sea going craft and providing it with gas turbine engines to produce compressed air held by the contouring skirt creates a hovercraft. The resulted greater pressure than the air on the upper side of the craft helped it to float and travel rapidly, because there is little friction with water or land. Adding fans to a blimp makes from it an airship. Adding a motor to a hand drill resulted in a power drill.

One of the newest in these assemblatory inventions is "**The Metabolic-Imaging Machine**" (MIM). Earlier "**The Positron Emission Tomography**" machine (P.E.T.) can reveal where tumor growth is taking place, while "**The Computerized Tomography**" scanner (C.T.) is used to show the precise anatomical structure. Seeing the data from C.T. and P.E.T. made at different times and with different settings in body positioning results in organ shifting. This usually resulted in difficult matching allowing for inaccurate interpretations. Now combining the anatomy of the examined part with the tumor growth as taking place by the "MIM" was for the hearts' ease of both patient and doctor. Thus, in revealing the precise location of a new tumor, enables the doctor to remove it before it metastasizes.

Such inventions, the **Assemblatory Inventions,** are here to stay. And because they have established industries, they will keep improving in shape and function.

The Conceptual Inventions

These are inventions triggered by simple phenomenons that are seen by many, yet they did not ring a bell except with the visionaries. So prophetics can see possibilities for such phenomenons to be employed for the benefit of mankind.

Examples of inventions from earlier times, which look simple yet were important landmarks for our ancestors and for our mushrooming civilization include:

Wheels:

Pushing a stone on a fallen tree trunk in ancient times triggered the invention of the wheel. Wheels are one of man's most useful inventions, because turning on an axle is a very good way to move loads. It was the Sumerians who made the first solid wheels from tree trunks five thousand years ago, while the first spoked wheels were made in 1600 B.C. by the Egyptians to be used in their chariots. Now we have grooved wheels for trains to run on rails, and metal wheels fitted with air-filled rubber tires for our cars, tractors, rigs, airplanes…etc.

Levers, Pulleys, and Tackles:

Using a rigid bar to lift a weight at one point of its length by the application of a force at a second and turning at a third on a fulcrum required the ingenuity of our predecessors to invent this simple lever. Nowadays, we rarely see levers in action but we rely more on their descendents, the pulleys and the tackles for hoisting heavier weights.

Pegs and Screws:

Pinning down materials with pegs resulted in the invention of a nail shaped piece with a spiral groove and a slotted head, designated to be inserted into materials by rotating with a screwdriver. Hence, the screw and the screwdriver came into existence. Usually the helical ribs of the screw achieve a better fastening of adjoint pieces than that resulted by using smooth, slender nails. These i:e the nails, were made by a nail making machine invented by Ezekiel Reed in 1786.

Examples of relatively recent inventions that were inferred by every day phenomenons that were seen by many, but were only made productive by the few gifted, comprise:

Taming the Steam:

Many have seen water vapor pushing the lid of a boiling kettle, yet it took ingenuity to use the power of steam to make the early steam engines. It was James Watt who built powerful engines in the early 1800's to power locomotives and steamships. They also powered the factory machines that made the **Industrial Revolution** of the late eighteenth century. The second generation was the steam turbines in which steam strikes cupped vanes attached to a shaft that turns at very high speed, now used in Atomic Plants, to generate about 75% of our electricity.

Entrapment of Sound:

Vibrations made by sound waves were used for a first time by Thomas Edison to make impressions with different depths on a tinfoil cylinder to record his voice. When the cylinder was played back, he was able to listen to his own voice. Edison's dog knew his master's voice, hence there was a brand of discs called, "His master's voice".

Preventing Dreaded Mine Explosions:

Working in their labs, researchers noticed that the flames from their Bunsen burners wouldn't ignite flammable gases when a wire mesh came in between them. This phenomenon was then used by Humphrey Davy in 1816 to construct **"The Safety Lamp"** to avoid explosions in mines. This was done by enclosing the flame of the lamp in a fine wire gauze from the surrounding atmosphere containing the flammable gas.

Wagdy A. Assawah

Lenses And Their Multi-usages:

Lenses were known to change the passage of light rays in different ways. They were and still are, used in glasses to correct near-sightedness or far-sightedness. The "burning glass" for producing fire from the sun's rays has been known since ancient times. The magnifying property of a single lens was recorded by Roger Bacon in 1200, yet it took ingenuity to arrange the lens to make microscopes to see the smallest of the small and to make telescopes to see the farthest of the very far.

A Blessed Mishap:

Wilhem Roentgen discovered X-rays by accident in 1895 while he was passing electricity between a cathode and an anode in a sealed glass tube. On a photographic plate lying under one of his books, that was exposed to the mysterious rays, he found a picture of his door key that was accidentally left between the book and the photographic plate. This observation could have been ignored and the value of using X-rays in the medical field could have been delayed. Yet, being a visionary, Roentgen knew its possibilities and now X-ray machines are widely used to examine defects in our skeletal system or to reveal the presence of foreign objects in the human body.

A Shrewd Observation:

Studying bacterial organisms grown in petri dishes, Alexander Fleming noticed the presence of an unwelcomed guest. The uninvited was a fungal colony of *Penicillium notatum,* which was killing his bacteria. He could have discarded the contaminated petri dish. Yet, from this phenomenon, which was encountered daily in micro-biological research labs, Fleming knew that the fungus must be exuding a mysterious substance capable of killing his bacteria. Could this substance kill also pathogenic bacteria in man? That was his question to himself. For that, he grew the fungus on a large scale. With the help of Howard Florey and Ernest Chain, these two doctors found a way to produce penicillin in large quantities. This new antibiotic was then widely used to cure infections in wounded soldiers in World War II.

Wagdy A. Assawah

A Messy Job Going Out of Use:

In the thirties of the departing century, for copy making of documents, people used to write the manuscripts with a pencil in which graphite was replaced by a violet lead dye, most probably made of gentian violet. The written document was then placed and pressed on a semi-solid gelatinous gel contained in a metal tray. As such the pressed document leaves a mirror image of the text that was written with the rich violet dye. When fresh pages, up to four or five, were pressed one at a time, on the gelatinous substratum, they will come out with an honest copy of the original text. Using a moistened cotton with a suitable solvent the traces left from the written material on the gel can be wiped out, so as the cleansed surface can be reused for another document replication. Responding to this human need, Chester Carlson in 1938 patented the first xerographic copier. Carlson had invented a dry process based on the ability of an electro-statically charged plate to attract powder in the image of the original document. The powder is then transferred to a piece of paper and a heater seals the powder to the paper. Hence, a copy emerge from the copier. Other copiers that can mix color to give a full color picture followed suit and also others that can make hundreds of copies with one push of a button.

Sterilization By Filteration:

Hundreds of years ago, masons used super-imposed sieves to sort out their mixtures used in masonry. Having different meshes, they were able to separate the pebbles from the grit, the grit from the sand, and the sand from the silt. This old methodology enticed the microbiologists to introduce a parallel for excluding the microbial contaminants from liquids that can be destroyed by autoclaving. Thus eventuating **"Sterilization by Filteration"** at room temperature. Other microbiological filters with different pore sizes are now available for separating yeasts from bacteria and bacteria from viruses. Consequently making the studies of such micro-organisms more effective on account of excluding the contaminants that can impair the results.

Conceptual Inventions Between Negation and Affirmation:

For a first impression, conceptual inventions will always be looked upon as absurdities. This is because the inventor is suggesting an unheard of solution for a contemporary problem based on an intuitive revelation of his. Hence the question asked will always be: "Is it a genuine presumption or is it a nonsensical impracticality?

103

Wagdy A. Assawah

Although appearing as illusory, conceptual inventions will only be appreciated by the insighted in the art. Once produced and noted as deliverers, they will be subject to improvements and metamorphosis. However, if there is an entrepreneur who shares with the inventor his visionary revelations, he will need new equipment, new assembly lines, and this in itself will be a set-back for new inventions to take off. Given also that a new product needs capital to venture with, the entrepreneur will find himself obligated to keep to the beaten track than to engage in a dubitancy. That is why, in the author's opinion, the concerned authorities should have a say in those critical conceptual inventions which can introduce beneficial modes in our lives. And even if submitted by an outsider of a particular field, they shouldn't be scoffed at. For once upon a time, a boy who spent only three months at school went on to lavishly produce over 100 inventions. The most famous, were the electric bulb and the phonograph. Had Thomas Edison been denied his creativity because he did not pursue his education, we could have been lagging behind, if not in a totally different world. There is great need for the insighted, knowledgeable specialists for refereeing and sorting out the worthy conceptual inventions from the reiterative inventions. And if dubiety rears its head, preliminary experiments should be carried out, and if the concept is proven futile, a negative result is still worth while knowing of.

THE FORWARDED CLASSIFICATION; IS IT A REWARDER?

Classifications and journeys have a lot in common. Both start with a first step. Unfolding both enrich our experiences by revealing new nooks and crannies to explore. We have seen earlier that the Periodic Table of the inanimate world predicted the presence of unknown elements. We also have seen that the evolutionary trees of the zoogenic creatures and phytonic beings offered sound explanations for the similarities or the dissimilarities within their different groups. Another surprising outcome was the presence of an uninterrupted continuity between the lifeless world of elements and the "lifeful" world of animals and plants. Now, do we hope to get comparable rewards from the here-in suggested classification? Though it may be rough at the edges, yet it has its merits. For that I'll leave this issue of being a rewarder, to the here-in extracted conclusions, hoping that they will settle the query and to justify the trial.

In the suggested distinctive four classes, one can find idiosyncratic peculiarities and shared characteristics that combine the inventions in each of them into an analogous assemblage. Better still, each class reveals in it's contrive the presence of a dominant

105

feature for a would-be patentee to follow suit to come up with a similar invention.

So for a **designer** who wants to improve an every day tool to replace it with a better one, he has to think, "Transfiguratory." Hence, if he gives an old tool a new twist by improving its appearance or its efficiency, he will be inventing a tool displaying a new feature.

And for a **renovator** who wants to improve an existing working machine in which a part is not doing its work properly, he has to think, "Amendatory". Hence, if he replaces a component with a better one, improving a machine's performance, he will get himself a patent.

And for an **innovator** who wants to come up with a dualistic gadget, he has to think "Assembleatory." Hence, by integrating a new element to a working contraption, the innovator gets a new dualistic machine and gets at the same time a well-earned patent.

But if he wants a home run and at the same time prove himself as **an insighted inventor**, he has to create a totally new, unprecedented instrument. And this will be, by seeing in an everyday phenomenon what everyone has seen, yet he has to think what nobody else has thought of. At first, the uninspired, be him a layman or a bookworm, will think that the inventor has come up with an absurdity, but when the conceptual gadget delivers, the inventor will be remembered in "The Annals Of History" for prodigally inventing a beneficial life-changing practicality.

From the above, the suggested here-in classification of inventions clearly demonstrates that inventiveness can be learned and that this classification can be credited for pointing out to would be patentees whether they are designers, renovators, innovators, or innate inventors how to take a dominant idiosyncratic peculiarity typifying a class and run with it. And for this classification to be a rewarder is a testimonial for its probity.

Yet another curious outcome that can be predicted from the above analysis, show that if the would be patentee wants riches, he better be targeting either the transfiguratory inventions or the amendatory inventions. In so doing, he will be allying himself with the industry that need not change its machines or its assembly lines, and consequently he will be earning his royalties with considerable ease.

But if the would be patentee targets the assemblatory inventions, and in particular, the conceptual inventions, he will be alienating himself from the humdrum industry. And he will be obligated to trust his invention to the future till someone, someday discovers the forgotten and misunderstood purport. Thence, the inventor will be getting his claim to fame and earning the praises when he needs them no more.

Saddening! Yet this is life. And life to rise above the colloquialism is a wicked discorder.

PART II

INVENTIVENESS AND THE INFLUX OF OUR MENTAL EXERTIONS

There are two kinds of truths: truths of reasoning and truths of fact.

Gottfried Wilhelm Leibniz

CHAPTER 3

INVENTIVENESS AND INTELLECTUALLISM

This chapter depicts an interdisciplinary attempt discussing an assortment of mental exertions, evinced in our humanitarian culture by those who came to this world before us. Though the given views mentioned under "Inventiveness and Mythopoeia" lack logic, yet they constituted a landmark on the road of our journey in time. Other contributed thoughts presented here-in stress the importance of theological convictions, science fiction, and intuitive deliberations. These intellectual pursuits continue to affect our gusto for inventiveness to procreate a better future.

I. INVENTIVENESS AND MYTHOPOEIA

The presented here-in doesn't concern itself with extraordinary deeds performed by gods, demigods, or legendary heroes of Greeks, Romans, or any other group or culture. What concerns us here is the invented myths that were circulated in past eras and which were neither apparent to the senses nor obvious to the intelligence. Peoples of earlier times, unappreciative of the systematized scientific approach, used to perpetuate unfounded notions around certain things to make them come in sight. Perhaps there were more of these mythopoetical beliefs that lack factual validity than meets the eye, yet antiquity chose to have them shrouded because of their absurdity. In the following some of these myths that were hoarded up in a coffer and were left to obliviousness.

Mother Earth is Flat:

Unlike the religion dogma that "Eve" was created from Adam's rib, this was a mythopoetical belief generated by man that the world is flat. It was one of those things that went on and on,

111

Wagdy A. Assawah

and to which Europeans succumbed to in the 1800's though Greeks, Romans, and Arabs knew better.

The Philosophers' Stone:

This is one of the most well known mythopoetical beliefs that is still remembered in textbooks. It is an imaginary stone believed to have the power to transmitting baser metals into gold.

Geese Grew from Barnacles:

Barnacles are marine crustaceans with feathery appendages for gathering food, that are free swimming as larvae but permanently fixed to rocks or boat hulls as adults. The myth says that geese grow from the crustacean. And this was a popular belief in Europe for a long time.

Lambs Grew on Bushes:

People of earlier times believed at one time, that the fibers from which they make their woven materials were from sheared lambs growing on bushes planted in India and other far eastern countries

Toothache is caused by tooth worms:

The earliest explanation of the cause of toothache was a "tooth worm" that resided in the gums and ate the roots of the tooth. The concept of worms causing dental decay was accepted without question for centuries in western civilization until 1728, when it was refuted by France's Pierre Fauchard, the father of modern dentistry.

Origin Of The Banana Tree:

An unfounded belief was put forward to explain how the banana tree came into existence. Myth-makers created the idea that by planting a seed of date-palm *(Phoenix dactylifera)* in the bulbous starchy corm of taro *(Colocasia esculenta)* of the Arum family, a banana tree will ensue.

The Red Spots That Asked For Avenging The Crucifixion Of Christ:

Peoples of medieval Europe, upon seeing red spots on their stale bread, believed they were the blood of Christ asking for avenging his crucifixion. Hence they became frenzied and possessed to kill the infidels responsible for the act. So they go on a rampage killing the Jewish people on account of this unfortunate apparition. Also peoples of those times, in a ritual called, "**Stabbing of the Bread**" used to do so on purpose. After a suitable time, the stabbed breads exhibited the red spots. To show their alleged loyalty to the church, they go on doing their heinous yet unpunishable crimes. As the science of microbiology came out of age, these red spots were found to be none other than yeast colonies that develop from yeast spores carried by air currents. Also, similar red colonies could be the result of growths caused by a bacillus called *Bacterium Prodigiosum.* Given the favorable conditions, they produced their ominous colonies on the breads that were usually kept in warm humid places.

From the above, it is shown that the scientific methodology provided the means to differentiate between myths lacking the needed factual validity and the truths substantiated by organized scientific experimentation. Had the man who mythologized that bananas were due to planting a date palm seed in a taro corm, done the simple experiment, he could have known the fallacy of his unfounded notion.

The Mystifying Expiration of Man:

Or

Debating The Mythopoeic Big Sleep:

Though it is not a perfect fit in the topic at hand, it shouldn't be left out from treatising. With its roots in an obscure past, stems in our contemporaneous times, and branches reaching out for an unknown future, this "Expiration of Man," the paragon of this creation, was and still is worthwhile bringing to the senses. With that said, my following deliberations shouldn't be delimiting to others from procreating. However, before I venture my presumptions on this subject, allow me to demonstrate the value of intuitive tangibility and how it can advent a righteous conclusion.

Wagdy A. Assawah

In the early two decades of the last century, books devoted several pages on proving that Mother Earth is not flat but a globular entity. Forwarded criteria entailed; seeing smoke of a steam-ship coming from behind the horizon before seeing the ship itself, that the more we rise above the earth's surface the more extensible the horizon circle gets, and if we line up three equally tall poles, say one mile apart from each other, the one in the middle will stand taller due to earth's curvature. Associated with this particular experiment, it may be worth mentioning that it was also used by Arab mathematicians to calculate the circumference of earth with an error of few meters.

In spite of the earlier submitted proofs, a visionary needed only to look at the sky to notice that all celestial bodies are spherical* in shape and to intuitively conclude that Mother Earth is in no different form. The question was then, were the earlier decades void of intuitive visionaries to declare such inference? Of course not. They were only afraid of being stigmatized with blasphemy. And we all know how Galileo Galilei was punished by the Church when he declared that Earth was not the center of the Universe.

---

* Forming spherical entities by different materials is a known physical property generated by surface tension. So oil droplets suspended in water, and for that matter celestial bodies with molten interiors, suspended in space have to conform to this maxim. It was Sir Isaac Newton's Gravity that combined the Terrestrial with the Celestial. Now on the Eternal Stage of Creation "Surface Tension" came in sight. As adduced by the author, it is acting equably on Earth as in Heavens."

Harkening back to expirations of humans, man without veritable reasoning, mythologized that he is going to be resurrected. This was precipitated by autotheism that he must be more valuable than the insects he crushes under his feet. It was also precipitated by his egoism that his intelligence, dexterity, and scientific acumen, earns him a raising from beyond the grave, and to go to Heavens to live happily ever after. However, with death on the prowl, insects, worms, flowers, man, and what not will get digested by nature, the inventor of the "Recycling Procedure."

Now, how can one convince the readers, especially after bursting their vanity bubble, that beyond the "Great Equalizer" there will be no curtain calling after crossing the bar! And to do so without the accepted scientific methodologies will be an abomination. In fact, the earlier fruitless trials of telenecromancy after burial, leave for us no alternative except to rely at least for the time being, on the revelations of visionaries. With their intuitive involvement, they are able to offer tenable conclusions that can stand on equal footing with those obtained by serious experimentations. And in the incorporated here-in verses, besides eloquently saying it all, they were also surprisingly complying with the accepted step-wise methodologies of scientific investigation. Thereby, the presented herein is for the reader to discover for himself this amazing collinearity.

117

Wagdy A. Assawah

Strange, is it not? that the Myriads who*   1. Methodology

Before us pass'd the Door of Darkness through   2. Observation

Not one returns to tell us the Road   3. Results

Which to discover, we must travel too   4. Conclusion

To recapitulate: matter can neither be annihilated nor can it be regenerated. And **Man upon his Expiration** will be going for good to an unknown "Great Beyond". Meanwhile, his matter (i.e. molecules) together with other molecules from other decomposing organisms, will be recycled into other vivacious beings and into other defunct objects. They, i.e. the molecules, will be logged in the **"Book Of Genesis,"** "Belonging to none, save the current users".

---

* *Fitz Gerald's Lyricals—Translation of Omar Khayyam's poetry.*

II. INVENTIVENESS AND THEOLOGICAL CONVICTIONS:

Not only do we invent materialistic inventions, but also we invent committing spiritualistic convictions. These convictions accompanied us for thousands of years and we have to standardize our everyday actions according to their fundamentalistic laws.

A world correlated with spiritual dimensions has been known for at least 60,000 years. When the primitive man became aware of his existence, he knew that he has to be born, to go from childhood to manhood, to get older and eventually die. His fear of death was a frightening mystification that he has to confront helplessly. In His dreams and delirium the dead appeared to him. Thus he was convinced that there are spirits existing after death, ever present but unseen. Also, ethereal spirits were attributed to clouds, rocks, rivers, waves, trees, animals...etc. This animistic style of religion took on a progressive, more abstract associations. Later, the spirits were looked at as deities equated with the forces of the Universe. Thence, pantheons were built and dedicated to them, where they were worshipped and appeased. Thus we find in Ancient Egypt a goddess for birth (Anukis), a goddess for love (Hathor), a goddess for fertility (Amaunet), a goddess for the underworld (Ammut), a goddess for mortuary (Anubis), a sun god (Atum-Re), an earth god

119

(Geb)…etc. Similar deities were known for Akkadians, Sumarians, Canaanites, Phoenicians, Greeks, Romans, Celtics, Nordics, Germans, Swedish…etc.

Many of the Ancient Egyptian deities were adopted by the Greeks and were given Hellenistic names. So (Horus), who avenged his father Osiris and regained his kingdom from Seth his uncle, was given a Greek name, Harendotes. And Hathor, the goddess of love enjoyed great popularity in Greco-Roman culture and many elements in the make-up of the goddess Aphrodite were modeled on her Egyptian style. Also in about 334 BC, Alexander the Great came to visit the Pantheon of "Zeus-Ammon" at Siwa Oasis in northern Egypt to consult with the oracle and to get his blessings before going to war against Persia and other Asian countries.

In the Mycenaean age, about 1600 B.C., Rome seems largely to have borrowed deities from Greece and renamed them. So Zeus became Jupiter and Aphrodite was named Venus. So the Polytheism that was known in Ancient Egypt was adopted by other Mediterranean cultures and beyond. Also being **"Sons Of Gods"** was another Ancient Egyptian exertion. So, for enforcing their authorities, Pharaohs claimed to be sons of gods. Roman Emperors followed suit, and they also declared their divinity and that when they die, they will be in the company of the Gods. Whether Pharaohs or Emperors, their priests claimed that they were anointed by Gods. To return the favor of the bestowed divinity on them, these rulers of the past gave the priests carte blanch to profit from practicing their religious rituals to their subjects. So both rulers and clergymen found sanctuary in this symbiotic relationship.

However, there was one man and the only one I know of in history who dared to confront this unholy alliance. This man was Akhenaton. He dismantled the old gods and set up Aten (=Aton) the sun god as the only god. Consequently, he destroyed the excessive priesthood and ended the exploitation of his people by this clique of profiteers. Thus, Akhenaton rightfully can be credited for introducing the concept of a single creator god, eventually monotheism. After his death, the new ruler, Tutankhamen found refuge in the older symbiotic relationship with the priests. Akhenaton's monotheism was thence abolished, yet its central feature survived in the three major Middle Eastern religions: Judaism, Christianity and Islam. Yet to the dismay of both the historians and archeologists, this Pharaoh's achievements were desecrated and his portraits were intentionally disfigured. He was depicted with an elongated skull to insinuate that he was mentally retarded and with heavy hips to make a lesser man of him.

The invention of theological convictions, that stood the tests of time, came with a number of life shaping beliefs. These convictions insinuated that we are an important part of this creation. That we did not come to this world in vain, but to fulfill a predetermined destiny. And though the material of our bodies may come to an end, yet our souls are eternal and separate from our bodies upon death. And that is why in Asian religions, a greater populous believe in Metempsychosis, i.e. the rebirth of the soul into another body, either human or animal. On the other hand, in the Monotheistic Middle Eastern religions, the followers believe that whoever wrought them in his image will bring them back to life on **"Judgment Day."** On that day, this **"Supreme Power"** will be meting out the rewards for the benevolent and the punishments for the mischievous. And after purgatory, we then become fit to go to heaven where we will, eternally live happily ever after. However, just in case there is no curtain calling, the earthly authorities instituted laws to condemn and punish the wrong doers in their lifetime.

Oddly, in Medieval times, priests and laymen alike, believed that buying "Receipts for Consecratory Pardons" issued by the church will exempt them from the chagrins of purgatory. And that the "Pardons" would let them go directly to "Heaven" to meet with God on His own turf. Had this continued to be practiced to the present day, our wealthy "Corporate Executives," affording top dollars for their indulgence, will end up occupying the best seats in "Heaven's Arena" too. Meanwhile, the poor and the needy will have no business to die, but to continue to load their daily 16 tons and to get deeper in debt, and Saint Peter shouldn't call them, for they owe their souls to the "Company Store," or for that matter the "Corporation Enterprises."

In conclusion, **"Religious Convictions"** eventuated by man are carved in our psyche, proving that invented thoughts outlive by far any invented contraption. Religions continue to be authoritative from one generation to another, affecting our lives in every aspect. Now our contemporary religions are much centered on the home, the sacraments of family life, naming the babies, initiation rites in adolescence, marriages, and burials. To disavow their fundamentals or to act discordantly, will be a breach in the accepted and may put the doer on the wrong side of the decreed.

The above, though abridged, yet it gives a bird's eye view on the changes that took place in religious exertions. These changes show that we drifted from believing in spirits as exhibited by animistic religions, to believing in deities as exhibited by polytheistic religions, to accepting one **"Supreme Power"** as in monotheistic religions. Yet there are incessant questions to which answers are sought.

With all the theological vocations exhibited by the peoples of ancient Egypt, Judea, and the northern part of Arabia, one cannot help asking, do the inhabitants of these juxtaposed areas, who were prominently Semitic, have hereditary genes for strong religious inclinations? Why do we fall away from new religions disclosed by such people as Jim Jones, David Koresh, and other doomsday cult preachers? Is it because these pastors are less potent than their peers of earlier times? Or is it because we don't want to waste monies on new clerical congregations? Or is it because we don't want new Worship Houses that do not die, yet per contra keep accumulating wealths across the ages? Certainly the earlier apostles are still governing from beyond their graves. On the other hand, the recent ones will only be remembered by their falsities.

Why do we feel obligated to put aside our minds and to accept unconditionally those convictions that came to us from an apparitional misty past without questioning? Why do existing factions resort to brutality to force their opinions on others in this age of systematic reasoning? Merely electing to resolve our religious differences by impetuous violence usher the bankruptcy of those inherited convictions. Earlier, humanity obliterated the abhorrent **"Inquisition Tribunals,"** managed to stem out the tide of burning alleged witches at the stakes. So can our intellectualistic civility of the early 21 century bring to an end our bedlamite obsession for religious fanaticism!

Now, allow me to fictionalize on behalf of our contemporaneous generations, to place Akhenaton as the first candidate for cloning? In his earlier life, he hung the bell around the neck of a lurking beast. Hopefully, the clone will be able to extricate hate from the hearts of fighting religious factions and unite humanity around good will to all men.

III. INVENTIVENESS AND FICTION

Inventiveness and fiction are two faces of one coin. To be a better inventor, you have to be able to fictionalize. And if necessity is the mother of invention, fiction is its legitimate father and the provider of its sustenance to keep it thriving. In fact, fictionalizing and inventiveness are bound together as sunrise and shadow. And to account for our predilection to fiction, the author finds in an answer for a question to Bernard Shaw, a satisfactory resolve. The question was, "Why do you hate death?" Unhesitatingly, he said, "Not being there when it happens!" This futuristic curiosity, which we all share for fiction will always be there yearning to be satiated.

Presently it will be distasteful to write about fiction coming into our lives without paying tribute to the founding fathers of this tantalizing art. The fact is, many consider that the contributions of **Jules Verne** (1828-1905) the French writer entitled him to be called "The creator of science fiction." He published several novels, the most renowned were *"Twenty Thousand Leagues Under the Sea"*, in 1870 and *"Around the World in 80 Days"*, in 1873.

In his first novel, he took a leap into the future and developed his ominous underwater machine, "**The *Nautilus***" in the image of our modern submarines. His sailors were able to walk the sea-bed to gather the delicacies offered by the ocean and to harvest the riches that other men had lost to sea. They used for this purpose breathing apparatuses like the ones used nowadays by our scuba divers. His fictionalized gadgets were not connected to the mother ship via long hoses to provide air for breathing as in the traditional ways of bygone eras.

Another science fiction mogul, was the English writer Herbert George Wells (1866-1946). His principal works were *"The Time Machine"* in 1895, *"The Island of Dr. Moreau"* in 1896, *"The Invisible Man"* in 1897, *"The War of the Worlds"* in 1898, and a collection of short stories, later combined as *"The Short Stories of H.G. Wells"*.

When *"The War of the Worlds"* was produced by Orson Welles and aired on the radio in 1938 for a Halloween treat, it caused considerable panic in major cities of the U.S.A. The people believed that their Mother Earth was really invaded by aliens from outer space. The truthfulness of Wells' imagination resulting in this

Wagdy A. Assawah

powerful reaction was a testimonial of Wells' ingenuity and earned him the title, "The Shakespeare of Science Fiction."

The Fictionalized Feats of Our Ancestors:

Our predecessors' fictions were the driving force for inventing many splendid things. Few of those fictioned feats that our ancestors dreamt of and which resulted in numerous outreaching inventions include:

The flying carpets, which later mushroomed into a family of different flying machines.

The crystal ball, which evolved in TV,s DVD,s and other devices for sending and getting pictures from outer space and from other planets.

The shadow-graphy throwing shadows of dancers and actors on a screen enticed the invention of theaters and movies.

The diving bell first used for dredging up sponges led to a variety of inventions: i.e. diving suites, submarines, and submersibles for deep-sea research.

One may miss the connection between those old wishes expressing hidden desires and the recently achieved inventions. That is because those hidden desires had to wait for our technologies to come out of age to surface again. On the other hand, nowadays we have more daring and ambitious dreams. With our powerful industries and our sophisticated tools such as computers, laser technology, nanoengineering…etc., the fiction of today is tomorrow's ipso facto.

The Fictionalized Feats of Our Contemporaries:

Man has always fictionalized to move faster. On land we have cars, motorbikes, tramways, trains, and elevated trains. On water we have speedboats, hydrofoils, and hovercrafts. In air we have an assortment of flying machines and supersonic airplanes. Man has always fictionalized to escape the earth's gravity. Now we have space shuttles that enable us to walk on the moon, build space stations, and send probes to picture other planets in our solar system. We are now targeting deeper space, so we may find another suitable planet to live on after we have destroyed and squandered the resources of our Mother Earth.

Man has always fictionalized to make good use of Mother Earth's bounty. And now he has dams to control rivers' revenues and to produce electricity from their surging currents. He also invented the seismograph in order to know where to dig for oil and the sonar to picture the topography of the sea floor to locate sunken vessels and other objects.

Man has fictionalized to live longer. Now the pharmaceutical companies are providing us with all sorts of tablets to subdue pain, to cure diseases, and to perform better in bed. The relatively new bioengineering enterprises, apart from offering prostheses to the impaired, are now providing us with all sorts of transmogrifatory instruments to help replace human lenses lost to cataracts with artificial lenses, to replace opaque corneas with transparent ones, to transplant kidneys, livers, and hearts, or to reattach severed organs and to examine blood vessels and other organs with the help of endoscopes without cutting open the patient. The newest endeavor, the nanotechnology, is offering nanobots engineered to carry drugs or genes for repairing organs or for correcting genetic disorders.

Thus, one day our medications will be administered by nanobots showered on our skin instead of tablets *per os.*

Man fictionalized to see the tiniest of the tiny. Opticians gave him an assortment of microscopes and Physicists gave him the electron microscope to get better and detailed magnifications. He wanted to see the farthest of the very far and Opticians gave him binoculars and different types of telescopes. Meanwhile, Physicists gave him the Radio telescopes to detect radio waves coming from faraway galaxies. These radio telescopes discover unknown bodies in outer space and, make with the help of computers, images of these objects. Man wanted to hear better, so Technicians gave him the hearing aids that can fit comfortably in his outer ear. Also, they gave him loudspeakers to make sounds loud enough to be heard at distances.

From the above, it is shown that science fiction, not science per se, that leads our inventiveness. Hence, to keep with the pack, I say: "Fictionalize, my fellow inventor...fictionalize."

Wagdy A. Assawah
The Tutorial Skull-Cap:

One may ask, "You advised your fellow inventors to fictionalize to keep with the pack! Do you practice what you preach?"

The **Tutorial Skull-Cap*** (T.S.C.) is a fictitious concept of the author. It was and still is an imaginary wishful thinking. However, our competitive edge technologies can transform it from the realm of the fictitious to the realm of the factual. Nowadays, what one man can dream, another man can achieve.

In my boyhood years, in the early thirties of the preceding century, I longed to have a fictitious head-gear that can tutor me into the subjects taught at school. To bring my concocted dream in sight to the reader, the envisioned head-gear could be called, "The Tutorial Skull-Cap." However, at that time, what I imagined couldn't

---

* *A personal thought from the typist: Typing for my father-in-law, I was astonished to know about his childhood-contrived T.S.C. Its likeness to the fabled, "Thinking Cap" that we were ordered to put on by our schoolteachers, cannot be mistaken. This indicated to me that it was a shared dream felt by youngsters and teachers alike in many countries. And that it is going to be an endeared universal commodity for eager youths seeking encyclopedic cognizances.*

be made into a working design. And for the best of intentions, it could be dismissed by the contemporaries as wishful thinking of a youngster. That is because the computer technology was a fetus yet to be born and the visual reality did not come out of age. Of course, my childhood T.S.C. was a modest apparatus intended to reiterate classroom lessons. But the T.S.C. that I am now suggesting is a sophisticated headgear that can display knowledges from all endeavors at the user's whim.

With the help of our expanding technological dexterities and our ability to store voluminous encyclopedic cognizances in minute gadgetries, the T.S.C. can be custom made. Subsequently, I was tempted to disclose my imagined contrivance that continued to pester my fantasy for nearly 65 years. My altruistic intention is to draw the attention of our computer savvy to build such T.S.C., and times are ripe for that. So our zealous students can put them on, whenever and wherever they like, to review their lessons or lecturers. So, Messrs William Gates, I call upon you to embrace this enterprise, hopefully it will give you more leverage for your legendary model of philanthropy.

Making it through, this suggested invention will smite the traditional school system and will be setting the stage for the open

Wagdy A. Assawah

school venture. Thenceforth, everyone will be able to study the combination of subjects he likes best without feeling pressured by the desires of others. Also, the early use of the T.S.C.s by youngsters will boost "as proven by the findings of neurologists" an early foundation of strong neuron circuitries in their brains, increasing the learning abilities of our coming generations.

IV. INVENTIVENESS AND INTUITION

While fiction is a passion of societies expressing passive wishful thinking, intuition is a personal revelation offering a visionary theory that may or may not fit the rational thought. Like a foreign body inserted in the soma of accepted perimeters, the new theory will not be easily accepted, yet it will activate research, pushing like fiction, our scientific frontiers into new horizons. And that is why intuition is on a par with fiction, if not the best catalyzer for exploratory quests.

A. Zero, the Mathematical Hero:

The invention of the zero was a mathematical intuition. On its own, it denotes nothingness. However, when juxtaposed against a numeral, it changes its value. On the right, it increases the value of the numeral by folds of ten: So 1 becomes 10 and 10 becomes 100 and 100 becomes 1,000...etc. While on the left, preceded by a decimal point, it decreases the value of 1 by folds of 10, so 1 becomes 0.1 (one tenth) and 0.1 becomes 0.01 (one hundredth) and 0.01 becomes 0.001 (one thousandth)...etc.

135

Wagdy A. Assawah

When we examine the Arabic Numerals as projected by their inventors, we will find why the numerals were inscribed as such and why the elliptical configuration was chosen to signify the zero. Doing so, we will find that the wise mathematicians of yore have assigned to each number as many angles in its original configuration equaling to the value it denotes.

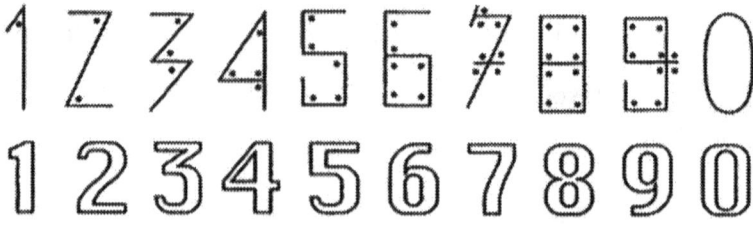

Fig. 2 Showing the original Arabic Numerals and the similized configurations

So the configuration denoting one has one angle. The configuration denoting two has two angles. The configuration denoting three has three angles, and so on till the configuration denoting nine having nine angles to recount its value. Thus, the diagrammatic representation given in Figure 2 was forwarded to elucidate this parallelism. Also to show that our mathematical hero,

136

"the zero" with zilch angles was an intuitive revelation to denote nullity.

And here is a teaser for you, my reader! Was this because we were given ten fingers? What, then, if we were given eight or six fingers instead of ten? Think about it! Someday we may have to communicate mathematically with aliens from outer space. Are they going to be bilaterally symmetrical or **radially triskelions**? And will their fingers be 6, 9, 12, or more? Shouldn't our mathematicians get ready to rise to the occasion?

B. The Continental Drift Dogma:

In 1922, Alfred Wegener, a meteorologist not a geologist, published his intuition suggesting how one supercontinent, which he called **Pangaea,** had split into huge chunks of land that drifted farther apart to form the continents we know of today. The reaction of established science ranged from ridicule to outraged rebuttal. Now we know for sure that the continents had once been one, and the world knew at last that Wegener had been right.

C. Administering Attenuated Microbes To Boost Immunity:

When Edward Jenner hit upon the notion of a smallpox vaccine in 1796, the Royal Society of London scolded him for risking his reputation on something "so much at variance with established knowledge, and withal so incredible". Now vaccination is a way for recurring lives and a way to eradicate diseases. We were practically successful in obliterating polio and we were nearly so with tuberculosis.

D. Clean Hands of Obstetricians; a Hygienic Must:

When the Hungarian physician Ignaz Semmelweis figured out that physicians' unwashed hands were causing fatal infections among new mothers at the University of Vienna in 1850, he lost his own position there.

E. Revealing a Microbial Connection to Ulceration:

Though Barry Marshall first reported his finding on the infectious cause of ulcers in 1983, his peers ignored the discovery until 1990, when the *National Enquirer* got hold of the story and told the world.

F. Suspecting A Microbial Cause for Heart Attacks:

On a comparable basis of finding an infectious cause of stomach ulceration, the author suspects that thrombi blocking the coronary arteries could have been triggered by bacterial nuclei. Engulfed by neutrophills to put them out of commission, they settle on the inner lining of the artery allowing a build up of a thrombus. Thus narrowed, the heart becomes deprived from oxygen, causing symptoms from tightness in the chest, to angina to, heart attack.

Corroborating this intuition, the occurrence of lesser cases of heart attacks in individuals accustomed to administer an occasional antibiotic every now and then. Yet as far as the author is aware, the litmus test that can settle this uncertainty has never been carried out.

Simple as it is, microscopical examinations of samples taken from these thrombi will provide a decisive answer and settle this query for good. Clearing this suspicion, whatever the outcome may be, will implement our views on heart attacks and may help in pointing out better ways to treat them.

G. A Potentiality For Curing Skin Cancer:

The northern shores of Africa belong ecologically and ethnically to the Mediterranean Corridor. If not nowadays, at least this was so in past eras before the on setting of desertification which is now juxtaposed to its shores. Enforcing this notion was and still is the ease with which to gallivant the Mediterranean and to intermingle with the people of the north than to cross the unsurpassable dry Sahara Desert to intermingle with the people of sub-Saharan Africa.

While the people of Northern Europe have relatively thicker, rough white skins to match their environmental conditions and to tolerate the winter subzero temperatures, those living in the Mediterranean Corridor have thinner, smoother olive skins. This difference, made by the presence of extra melanin, was a blessed adaptation concerning the Mediterranean inhabitants living on both sides of this basin. However, this difference in melanin content may also be attributed to an evolutionary process known to geneticists as "Reversion to Type". This extra melanin, as supported by scientific findings, helped in curbing both the harmful effect of the fierce summer sun and the ravages of the ultraviolet rays.

Whether this difference in melanin content was due to adaptation or due to reversion to type, it certainly provides the olive-skinned people with a free protective sunscreen tan. This tan kept at bay a variety of epidermal disorders and other malignant skin cancers induced by the ultraviolet rays that can metastasize and spread through out the body. For understanding the gravity of such pestilences, a gracious lady, **Maureen Reagan**, the daughter of former President Ronald Reagan, died at the age of 60 in August 2001 from a malignant melanoma diagnosed in 1996, which progressed to tumors in her hips and brain.

Now with our awareness of the beneficial effect of the melanin pigment, can we put it to work in our favor? Certainly the people who have it were not given it to boast a Hollywood tan over the Casper white souls! Thence, the question is, do the melanin-rich cells produce a "Nullifier" to stop the cells at the base of the epidermis from becoming cancerous? This presumption is definitely in need of tentative systematic research. And if such a "Nullifying agent" is confirmed as a participator in the battle of skin diseases, then biochemists can extract it, identify it, synthesize it, and concentrate it to increase its potency in tests. And when tests prove its efficacy, we will be having a new medication to cure the afflicted from the nightmarish skin cancers without the dangerous side effects caused by the nowadays used antineoplastic drugs.

H. DNA Recombinants Cogenting Evolution:

The forwarded here is an earnest request to decipher evolution at the genetic level. Earlier studies pertaining to the evolutionary dogma, whether in plants or in animals, were concerned with the changes in morphological features, in structural differences, and in comparative anatomical details. Should biologists then stop researching because the earlier classification systems have done their good deeds? Of course not. It is time for biologists to push the envelope further to widen our frontiers abutting the unknown. Surprisingly, the numbers of genes in different living beings from *drosophila* to man are of no big significance. This fact left geneticists wondering about where the weighty differences that resulted in shaping man, the paragon of creation, the ways he is!

Thence, evolution as such, cannot be attributed only to the number of genes, but also to the wondrous proteins they spin-off. Such different proteins are the building blocks that are responsible in shaping creatures differently and by investigating them, we may be deciphering the language in which nature has created life.

I. The Fallacy That Humans Are Racially Different:

Yet, all intuitions are not born equal. The following is a singled example of a once-applauded idea which latter went belly up. In contrast to the earlier intuitions, the ***Origin Of Species*** forwarded by Charles Darwin came up with a **femme fatale** insinuating that humans are racially different. This imputed superiority was clangorous with the temperament of the Europeans of those times. Unfortunately, it caused **unwelcomed transgressions** to many, equaling if not in excess to those harms inflicted by fanatic religious factions across the ages.

Had the Europeans looked into the matter that created the alleged gap in civility between them and the unlucky nations of Asia and Africa who were branded as socially inferior, they could have spared the world a lot of destructive hatred and they could have ended

Wagdy A. Assawah

up knowing better. In fact the difference boils down to longevity in the lives of individuals on both sides of the issue. In Europe, better climates with minimal diseases, together with abundance in healthy sustenance favored longer lifespans. With healthy physique, people persevered, thus having time to get educated, gain experiences, and to contribute new revelations to be left behind for others to build on and to establish a growing culture. On the other hand, in the poorer nations of Asia and Africa, for no fault of their own, they were destined to have shorter lives. This was due to malnutrition and due to the abundance of diseases caused by hosts of viral, bacterial, and protozoan organisms. Hence, they were preordained to die young. So, the lucky few who managed to reach manhood had only time to get educated, but no time to gain experiences, let alone passing them on to future generations.

To add insult to injury, these nations were also plagued by colonization and exploitation by the Europeans. With wealth without toil and times without worry, the Europeans helped with the longevity in their lives, allowed the gap to grow wider and wider between the haves and the have nots. Nowadays, the so called retrogressive races are forming minorities in most of the European nations and in the American continents, and are proving themselves to be on a par with the natives of these continents.

Now the recent advancements in genetics, proved what we have already known, that there is no such thing as genetic racial distinctions, and that we are all one race with similarities that reach 99.99%.

Discussing the value of longevity and that thoughts and gained experiences can live and get passed from one man to another to enrich the human culture, brings to my mind the interest in horizontal interdisciplinary endeavors. Believing in the value of such undertaking, I deliberately engaged myself in weaving a harmonious mosaic about inventiveness of gadgetries, inventiveness of thought and other sciences and arts. However, such interdisciplinary attempts will continue to be the privilege of those who live long, escape Alzheimer's disease, and have the insight to forge from their life long cognizance new terrains.

V. INVENTIVENESS, THE CHRONOLOGIZED, AND THE CHRONICLED

A. The Chronologized from Earlier Cultures:

Apart from the theological beliefs, the great cultures of Ancient Egypt, Mesopotamia, Sumeria, Akkadia, and other adjacent lands in Asia Minor and greater Asia formulated the roots of our modern sciences. Their scientific primogenitors comprised: Mathematics, Astronomy, Chemistry, Engineering, Medical, and Pharmaceutical endeavors. Entrenched in those early cultures, they created an undoubted civility which one can rightfully call, "The Eastern Tradition." Picked up and contributed to by later societies living in countries bordering the shores of the Mediterranean Sea, our contemporary historians ignored the efforts of jump-starting those early endeavors by the earlier Eastern civilizations. And in the absence of logic, they claimed the whole gamut to be solely a Western Exertion, and hence they named this cultural continuity, "The Western Tradition." This was certainly unfair and does not concord with the historical actuality.

Undoubtedly, the contributions of the earlier "Eastern Provinces" that were rightfully called "**The Cradle of Civilization**" and which were later continued by other European countries, resulted in mushrooming cognizances beyond our expectations. Yet this does not justify ignoring the founding cultures. So please, Mr. Eugene Weber, stop calling your episodes, "The Western Tradition". We the viewers of your tantalizing exposé request you to call it "The Humanitarian Tradition." And though those traditions are still westward-bound and flourishing in America and Japan, we will never accept calling them either "The American Tradition," nor "The Japanese Tradition". We definitely want to acknowledge the efforts of all contributors, especially the founding cultures and rightfully call it "The Humanitarian Tradition". This is especially so, since our technologies are making from our world a "Global Village."

B. The Chronicled in Our Contemporary Dystopian Culture:

Intuitive philosophers dreamt from times before time of idealistic communities in which perfectibility of human life is culminant. In his book, *The Republic,* Plato (427-347 BC) described his idea of a perfect nation. In his 1516 *Utopia,* Sir Thomas More imagined a place of ideal laws, government, and social conditions.

147

Wagdy A. Assawah

While in his 1933 *Lost Horizon,* James Hilton advocated an imaginary place called "Shangri-La" where life approaches perfection.

Towards Establishing a Global Moral Authority:

With the dawning of the 21st Century, the world is unconsciously dancing on a crater of a seething volcano. Unfortunately, this situation is created by mistrust in each other, by the availability of mass destruction technologies in trigger happy hands of fiendish, psychoneurotic leaders. Thus, it is no prophecy that if we don't wise up, we are going to experience patchy grinding wars that will eventually trammel in all nations. However, history taught us that dire situations are usually matched by equally determined and resourceful sagacious men. It is now that our leaders should rise to the occasion, come together to stop this evil in its infancy. To hope not for perfect communities, or for idealistic nations, but for nothing less than "**Utopian Global Coexistence**". The fact that our U.N. can incarcerate criminals of ethnic cleansing and send them to the U.N. Tribunal at The Hague to be condemned for their heinous acts is an indicator that we are stepping in the right direction.

For that, kudos go to our U.N. Secretary General Kofi Annan for restoring the lost positivity of our U.N. He is now adopting another heroic plan to combat the pandemic AIDS, and this is another achievement under his belt. Now, if we want for our U.N., "which came into being in 1946," to continue with positivity as the supreme guardian of justice and liberty for all nations, then we must bolster its capabilities. Otherwise, if we don't, it will be another lethean organization like its predecessor, "The League of Nations" first established in 1919.

Hence, for letting our precious U.N. do what no other organization has done before, and stand for something that the world honor and admire, we should for our civility's sake, insist upon the followings:

• Renounce the veto* privilege enjoyed by the powerful nations. For a reason, it does not sit well with the spirit of democracy we proudly claim that we are. For another, humanity looks upon such illustrious nations as "Big Brothers" and not as high technocratic bullies that want to dominate the world.

---

* We "The Americans" speak volume of our benevolent democracy. Enthusiastically, we don't mind to go the extra mile, to establish its bounty in other countries. However, at the U.N. level, "The Veto Clique" dislikes establishing democracy among Nations of our circumscribed Globe. Irreconcilably the V.C., with a straight face declares **"We Are Even Handed"**.

- No country whatever its status may be, neither use its supremacy to manipulate the U.N. to do its bidding nor to jeopardize its existence. This congregation of hundreds of world leaders deserves not to be scoffed at by the mighty nations. Reaching a resolve for a dilemma scrutinized by multifarious intellectuals will certainly be more righteous than a resolve adduced by one nation driven by its extraneous ends.

Earlier, such schism between the powerful and the powerless brought The League of Nations to its demise. It is therefore unwise not to learn from our mistakes and to let the same inequitable fate befalls the conscience of our world, i.e. "The U.N.." Letting this to happen commits us to the rules of the jungle. Thereby, we will be retrogressing to hordes of bipedal primates that act on whims and not on reasoned ratiocinations.

- Muster **"An International Striking Force"** recruited from all nations, equipped by the most overwrought appurtenances in our arsenals. Acting under the auspices of **"A Global Military Umpirage,"** this force is meant to go and separates the warring factions, to arbitrate and to come up with a fair resolution. This latter being exacted and accepted will hopefully make an aggressive military incursion by the International Striking Force an unneeded disciplinary operation. Yet if one opponent takes a turn to the worse, he will be mutinying against our worldwide solidarity, hence bringing upon himself the wrath of all nations, together with other punishments as decreed by our international laws.

- Levying annual assessments to be forwarded by all nations comparable to their financial assets, so that the U.N. organization will have its own resources to execute its varied responsibilities. Few of those money-consuming assignments include: fighting global poverty and hunger, protecting endangered species, preventing the deterioration of established ecological systems, fighting global warming and the destruction of the ozone layer, protecting archaeological sites and stopping of smuggled antiquities, brandishing the new slavery rampant in certain African nations catalyzed by the lust for precious stones and catalyzed by other lucrative agricultural products such as cocoa beans, cotton, sugar cane…etc.

- To entice all nations around the globe to amend their constitutions with laws to stem the tide of hate crimes. And to specify the punishments that fit the acts of discriminating against race, ethnicity, religion, sexual preference…etc. Hopefully this forwarded suggestion will make a better place of our tumultuous world, since it will usher to the leaders of rogue nations that the times for **Neronian** acts went for no return. Thus, I appeal to our U.N. under the auspices of our illustrious **Kofi** to adopt this deliberation. Thus achieved, our U.N. and the leading democracies will hopefully need not deploy troops to prevent the heinous genocide in other countries. Since establishing such laws addressing hate crimes on a worldwide level, will on one hand provide the discriminated against with the legitimate grounds to take the perpetrators to a U.N. court of justice. And on the other hand, it will obligate a high-tech bully to think twice before going on a killing spree to eliminate a co-habitant minority.

- Because the powerful rich nations are interested in star-wars, outer-space exploration and their daily worries, the U.N. should create a corporation for adopting peace promoting technologies to help the peoples of under-developed countries to live healthier and happier.

Unless we do all of the above and then some, humanity, especially in the poorer nations, will be jeopardized and the ripple effect sooner than later will catch up with us. What I am arguing is that this U.N. shall never perish from earth, to insure that our benevolence will always match our civility. And to insure that violence and injustice rife on every side will be hampered for good.

CHAPTER 4

INVENTIVENESS AND THE TABOOED

Up till now, our inventions went hand in hand with our legitimate necessities. However, our strive doesn't warrant us to act against the perimeters of our creation, even if we have the means to do so. Disturbing the balanced axioms developed by the Universal Wisdom will open **Pandora's boxes** for unseen abominations. Also with man's mixing hate with his unlimited abilities to create destructive weapons, frightens the dickens out of the wise. This is particularly so because man is still exhibiting more brawn than brains.

In my humble opinion, re-creating ourselves by genetic engineering or eventuating ways to negate nature's doctrines are acts of arrogance. And we ought not to grumble when the belated

153

punishments exacted by her, fit our recalcitrancy. Also Mother Nature neither wants us to travel in time, whether by a man made machine or by cryogenics. And here is why!

I. CONCERNING CLONING

The following work was granted a certificate of registration from the U.S. Copyright Office. It was submitted under the title:

"CLONEES, A PRODROMAL TRAJECTORY"

Registration number, TXu 818-893   Date of Registration 09/03/97

Cloning was known in plants before it was used to clone animals. Grafting a bud is cloning. Planting a nodal stem-piece is cloning. Certain cacti dropping small sprouts or whole leaves in nearby soils to root and give rise to new plants, is self-cloning. Cloning of plants; a gratuity from Mother Nature, is a heavenly brew. Needless to say, plants provide us with food, medicine, shelter, clothing…etc., for humans and animals all alike.

On the other hand, propagating humans from mature cells is a method that was not bestowed on us by Mother Nature, is a witch's brew.

We have busted the seams! Do we really want more on the ark? That is the question? And the accepted answer will only be from those specialists walking in the boots of Thomas Robert Malthus (1766-1834). The author of *Essay on the Principle of Population,* in which he suggested voluntarily controlling births. In essence, our humanity ask only for those who come in response to human chemistry and add variety to the populous, and not for those who come in response to human greed and add only homologous replicates of Cains assembling other Cains.

The technology of human cloning, or human xerography as it is sometimes called, involves simple methodology. With the help of a micromanipulator annexed to the stage of a microscope and an ultra-fine micropipette, it will only take few hit and miss trials to master the technique of cloning. The fortune hunters of private sectors and their hired clonomaniacs, will sing the praises for the benefits of cloning and they will do their best to legalize the

Wagdy A. Assawah

technology for their own extraneous ends. And consider the cloners who will sell their services to the terrorist groups!

Cloning coming into our lives poses ethical and scientific questions. Both geneticists and physicians knew about the outcomes of cloning. In the majority of trials, their experiments on animals showed that cloning will either end up with dead embryos or in deformed specimens. Similar results and deformities are bound to happen in human cloning, producing pitiful specimens. However, if success emanates, the resulting clone, with no family to own him, will continue to suffer from a "disownment syndrome". Other community problems, comprise circumstances that will obstruct justice such as catching a thief, while all clones have the same facial features and the same fingerprints. Or trying to convict a rapist while all the clones have the same DNA pattern…etc.

However, vilifying cloning without forwarding strong reasons is unacceptable. Apart from the ethical anxieties which I shall leave to religious factions, to social artisans, and to forensic experts to entertain, there is scientifically more wickedness in cloning than meets the eye.

A. Sporadic Intelligence Versus Gregarious Intelligence:

There are considerable differences between sporadic individual intelligence as evolved by natural selection and gregarious group intelligence as manufactured artificially by cloning technologies. Sporadic individual intelligence cherishes his own achievements and will be contented with the recognition he is worthy of from his peers and his country. On the other hand, gregarious intelligence takes pleasure in dominance and in forcing every activity to circumgyrate around his own interests. Many groups of humans, believing that they were a cut above the others, were able to inflict holocausts and grinding wars on mankind.

B. The Susceptibility of clones to superiority complex:

As per their temperamental makeup, clones will be forced to appreciate their own kind. Compared to them, earthlings will be mediocre, weak, vulnerable to diseases, subject to variation in traits, and victims of genetical disorders due to sexual reproduction. Thus, earthlings will be pitiful creatures unworthy of inheriting the earth. Eventually, driven by their superiority and maniacal egos, they will

157

decide to waste the earthlings, the unwanted extra load on the ark. And this is an impending possibility.

C. The Bio-disasters that come with Alien Intruders:

Introduced intruders in a new habitat usually upset its balance, resulting in its destruction. What then, if our new intruders, over and above their multiplicity, are the intelligent, resourceful clones! Ironically, we don't lack admonitory examples. The Conquistadors annihilated the Mayans, the Aztecs, and others living in Central America, reducing their cities to smithereens and committing their civilizations to oblivion. The greedy European settlers massacred the Red Indians of North America to near extinction and kept the remaining few in settlements.

Will the clones spare us a lesser fate? It will be naivety to expect from them, compassion or benevolence. And this will be like expecting honey from the fangs of a viper.

D. Cloning, a Genetic Bomb of Our Own Making:

Unlike atomic bombs, our man-made genetic bomb does not lack neither the brains nor the cunning to auto-inflict their wrath on earthlings. We their "inventors," have provided them with mega-genes that display every human trait at its fullest capacity. Also with the passage of time, they will perfect their rivalry edge and emblaze their innate viciousness.

"The Planet of Clones" will now be in full regalia and the earthlings will be programmed with the help of implanted microships to do the biddings of their masters.

Unlike the Atomic Bomb, which can be outlawed internationally, the microbes to which we can develop immunity, the eruptions of volcanoes, the disasters of floods, tidal waves, tornadoes, droughts that can be predicted, ameliorated or avoided, unlike the asteroids that can be intercepted while still in outer space and diverted, the clones our invented "Genetic Bomb", will be an unstoppable anathema.

Wagdy A. Assawah

So, on behalf of the coming generations, I supplicate to our leaders:

1. Leave intelligence, healthy physique, beauty, and all other human traits to evolve at their own pace.

2. Do not let humans to be replicas of one another. There is sagacity in diversity. Assortment of creatures is what keeps the different components cooperating harmoniously.

3. Our humanity will be much happier if our prestigious scientists devote their timal and efforts in genetical engineering and transgenic animal experiment to elevate human miseries and famine. In conclusion, cloning is a prodromal trajectory and I steadfastly believe that clonocopying of humans should be tabooed.

II. EVENTUATING UNNATURAL SEXUAL PREFERENCES BY MAN

One of our universe's rules asserts that "Opposites Concord" while "Analogues Discord". Examples supporting this rule are numerous.

In magnetism, different poles concord, while similar poles repel each other, i.e. discord. Actually the coherence of a piece of iron is due to the alignment of its moieties with the negative ends consorting with the positive ends and vice versa.

An electric current that lights our homes and runs our factories will not deliver unless it flows between two concording opposites. So to do its job, it has to flow from a positive outlet and goes back into a negative inlet.

From the limitless macrocosms where our vulnerable earth is a speck of dust, to the microcosms of the ultramicroscopic atoms, the opposites whether suns and planets as manifested in the macrocosm

Wagdy A. Assawah

or protons and electrons as manifested in the microcosm, are meant to concord to keep these systems functioning.

Life in the plant kingdom cannot continue unless the process of pollination and fertilization is consummated between two sexually different conjugating gametes. In fact, all creatures from man to animals, birds, insects…etc., owe their existence, generation after generation, to the bounty of the different products rendered by the sexuality displayed by the different plant species.

Humans, animals, birds, fish, worms, insects, you name it, will come to an end unless mating takes place between two opposite yet concordant sexes. It is the coming together of a spermatozoon and an ovum, that guarantees the continuation of life on earth. Hence, eventuating homosexuality between analogous sexes is against the accepted doctrine of Mother Nature. Consequently, the aspect of homosexuality neither gives us the excuse to practice it nor to declare it to be on equal footing with the sexual phenomena credited for keeping life going on successfully till now.

Our living together in large numbers, is in itself a hazardous sociability encouraging the spread of diseases. Examples corroborating this include the endemic occurrences of cholera, dysentery, influenza…etc. in over populated areas. The pandemic occurrence of bubonic plague in Europe in the middle ages and the spread of HIV in sub-Saharan Africa in our contemporaneous times. And now to add to our congestion in metropolises, our bodily contacts, will provide a Trojan entry of contagions in redoubted bodies. Unfortunately, our payback for a fleeting moment of forbidden pleasure will be venereal diseases that we know of and other maladies that we don't know of, which may be impossible to cure or to stop from spreading to the innocent.

And what has been forwarded here is built on scientific grounds and not on a reiterative sermon delivered from the pulpit.

III. EVENTUATING CANNIBALISM IN LIVE-STOCKS

Live-stock has lived since the dawn of history, feeding on hay, fodder, and other plant products. Obligating these herbivores to feed on remnants of other animals for rushing results is really forced cannibalism on these creatures and does not tally with Mother Nature's calculations. Unfortunately, this resulted in an unheard of disease of cows now known to us as the "Mad Cow Disease." And to add insult to injury, this disease is easily transmitted to healthy humans upon eating contaminated meat from diseased cows. The afflicted experience agonizing pains, loss of memory, and loss of motor coordination, and gets wasted in a matter of weeks to few months. Examining specimens of their brains revealed the presence of moribund patches caused by contagions known as "prions".

Because of the patchy symptoms in the infected brain tissue of the patient, the disease was called, **"Bovine Spongiform Encephalopathy"** (B.S.E.). The disturbing fact about this disease is that up till now, these prions are indestructible. They can stand freezing, deadly chemicals, such as formaldehyde, and extreme heat, and come out unscathed.

Later studies on these prions proved that they were related to a similar disease infecting sheep, called, **"Scrapie"** known in Europe since 1730. The recently-adopted practice of feeding cows with the remnants of other farm animals, including sheep, caused the contagion of scrapie to mutate in cows and to be capable of infecting them with a new variant of prions. Unfortunately, the cows acted as a bridging juncture for these newly evolved prions to jump the gap from cows to humans. Earlier people used to eat sheep infected with scrapie without getting the disease.

However, before scientists knew that contaminated animal feed to be responsible for the spread of mad-cow disease, Mother Nature portended humanity of a horrifying potentiality. Her message was cloaked in the parable of a Swedish cat called "Bits". It died more than seventeen years ago, after falling ill in 1985 from pet food that contained pulverized meat and bone meal from ground-up cat and dog carcasses. And this was a forewarning to Agricultural Sectors and Animal Feed Industries. Unfortunately, the messenger was a frigid iceberg showing Bits the cat for a tip. Naïve, the addressee trivialized the admonition. Bits' reaction to canned food became so severe that it licked away all the fur from her tail and hindquarters. The pet-food scandal resulted in a 1986 governmental ban on meat and bone meal in cattle fodder. Had other nations at this point in time stopped feeding cows with fodder containing animal remnants, the story of B.S.E. could have been different.

Likewise, it may be worthwhile to point out that there was a similar disease to B.S.E. in certain tribes that practice anthropophagi. Those tribes were discovered in the forests of New Guinea by German explorers in the mid of the nineteenth century. They used to eat their dead out of respect, eventuating a fatal brain disease known as **Kuru**.

Wagdy A. Assawah

In conclusion, inventing a new fodder tainted with sheep offals and obligating herbivores to become carnivores was negating one of nature's doctrines, and we are now living to regret it.

IV. CONCERNING FRAUDULENT PROPAGANDA

There are many unjustified propaganda aimed at promoting certain products. Unfortunately, the industry leans heavily on graphic artists to substantiate their allegations by producing animations supporting their skewed views. In so doing, they are using the dexterity of the artists and the sentient notion idling in our psyche that a picture, or for that matter an animation, is better than a thousand words. Ganging to trivialize the intelligence of viewers by the industry, by the co-operating graphic artists, and by the TV stations, should be looked into so that only the rightful will be allowed to surface. Though man-made laws do not protect the callow, yet ethics should compel us to taboo fraudulent advertisements.

Another betrayal of the trust comparable to the stealing of one's identity rampant nowadays will be plagiarizing the works of others. No doubt, there are considerable numbers of intellectual elderlies who were unable to ride the wave of the mushrooming computer gadgetrie. Unable to catch up with the mushrooming changes, they will be obligated to trust the computer-savvy with their contributions in different endeavors. Leaving this world before their brain-children were divulged, will entice the bad seeds to attribute such left-behind works for themselves and to take credit for toil that was never theirs and insight that was never enjoyed by them.

V. "DEEDS" SPONSORING RESEARCH TERRITORIES! AN UNNEEDED IGNOMINY

Dissimilarities abound between researching and inventing, holding them diagonally opposed. Consequently, this makes what is relevant for researching irrelevant for inventing.

Scientists do their research in laboratories at universities, at pharmaceutical companies, at medical schools, or at other diversified research institutes. Meanwhile, inventors do their inventions in workshops or in factories built for the purpose.

Choosing to devote their lives to promote human conditions, scientists are paid by their institutions, though a drop in the bucket, to do their noble work. Meanwhile, inventors have to gamble with their capital on whatever they intend to come up with.

In biology, bioengineering, medical research, pharmaceutical medicaments, agriculture endeavors, animal husbandry...etc., researchers publish their scientific findings in comparable scientific journals. And they are privileged to go down in scientific publications and in textbooks as renowned scientists for their achievements. Meanwhile, inventors have to patent their inventions with the concerned authorities, and if their gadgetries appeal to the industry, they will be getting their belated earnings as royalties.

From the above, it is shown that scientists researching in different endeavors have committed themselves to a noble cause. This commitment is comparable to the oath of allegiance to one's country. Yet the point here is not for denying them their well-earned share in the created wealth, but to blame the system that allowed our scientists to live on the verge of financial insufficiency. Essentially, the point here is to stop giving "deeds" i.e. patents for territorial research with the purpose of preventing other scientists from tackling, though from other points of view, such sponsored terrains. Because in so doing, we will be interfering with the correctness of deduced conclusions. For in researching, conclusions are continuously questioned, and in time, corrected. Henceforth, if we allow the "deed" predilection to be an accepted norm, we will be retrogressing in time to where cognizances were obscured from the populous; meanwhile, they were sacerdotal to the **"High Priests"** of the dark ages.

- Imagine an astronomer that was given a "deed" to monopolize a black-hole that was discovered earlier by him, and preventing others from studying other various phenomena associated with this black hole!

- Imagine a microbiologist that was given a "deed" for discovering a bacterium that can digest oil lost to the sea, and preventing others from working on mutants of this bacterium that can be more efficient in devouring the oil slick in a shorter time!

- Imagine a geneticist that was given a "deed" for introducing a new lucrative crop, preventing other researchers from improving its resistance to insect attacks and improving its immunity to phytopathogenic organisms!

- Imagine a cardiologist that was given a "deed" for discovering a new venue for feeding the muscles of the heart with oxygenated blood, and preventing other cardiologists from making use of his new method except through his expertise!

In conclusion, don't you agree with me that all of the above belongs to humanity with no strings attached?. Obstructing the march of our scientific endeavors by giving "deeds" to researchers and scientists to protect patches in the scientific terrain, is not in favor of a healthy scientific climate. And besides inducing perplexity about what is covered by a "deed" and what is not, the practice doesn't concur with our humanitarian traditions. Thence, this "**Mega Taboo**" should be stopped in its infancy before this gaping schism overwhelm the menders.

VI. PATENTING THE ILLUSORY! AN UNWELCOME INTENTION

Submitting an invention for patenting necessitates the presence of endowed prerequisites in it, to earn a patent. The most important of these prerequisites are:

A.   Detailing the parts making up this particular invention.

B.   Explaining how these parts fit snugly with each other.

C.   Showing how the assembled parts work harmoniously to achieve an aspired intent.

D.   And most importantly, how others can build it, work it and use it for the purpose contemplated.

Consequently, on account of lacking one or all of the endowed prerequisites mentioned above, one cannot allow patenting the fictitious "Crystal Ball" of early storytellers. And this is enforced too, by the fact that this crystal ball cannot show us our loved ones in a far away country. Neither can we allow patenting the "Flying Carpet," and this is especially so because it will fail to take us across vast areas to visit with other nations. Also, one cannot patent spaceships claimed to be seen by many across the ages visiting our planet. Yet this chimera depicting a spaceship envisioned by our apparitional ancestors of bygone eras, was submitted and was granted a patent, in spite of lacking all of the above mentioned prerequisites. It also ignored an essentiality regarding that it could be rebuilt by others and to do what it claimed to do, i.e., the ability to explore outer space. Consequently, and for the sake of doing the righteous, it is a taboo to patent a semblance to make of it a factuality.

VII. CONCERNING TIME TRAVELING

A. The Earliest Envisions:

Time traveling has its interesting history. In 1895, H.G. Wells wrote his fictional story, *"The Time Machine",* in which his valiant commuter journeyed horizontally to the past to relive it, and to the future to explore it. On the other hand, Dino Buzzati, an Italian writer who emerged after WWII chose "Diacosia City" for his story players to travel vertically in time to double their lifespans. In his fable, Professor Aldo Cristofari invented a special electrostatic field called "Field C" within which natural phenomenons required an abnormally longer period of time to complete their life cycle. This meant that an organism with an average life span of ten years could be inserted in "Field C" and reach an age of twenty years. Man could now hope to live for two centuries. And so, from the Greek for two hundred, the name "Diacosia" was chosen for Professor Cristofari's City. Hence, Diacosia became the greatest wonder of the world.

However, the visionary, Edwin Mediner published that "Field C" could operate in both directions and he wrote an article that proved to be the death knell for Diacosia. And one day "Field C" reversed its effect and the healthy young of Diacosia, in the vortex of unleashed time, became old with white hair, flaccid and bony. Then they collapsed, died, and decomposed into a white dust. Thus, Mediner's prophecy came true.

Wagdy A. Assawah

B. The Latest Ventures:

Time is the fourth dimension of space. It adds in the lengths of the three known coordinates, i.e. length, breadth and depth. Mother Nature doesn't want us to time-travel, neither into the past nor into the future. We are only permitted to travel in the three coordinates of space during the contemporaneous times of our existence on earth. As to time traveling into a folded away past, we are only allowed to regurgitate the histories we know of. As to time traveling into the future, we are only allowed to wait helplessly for the unknown to be revealed. In reality time traveling, though, it furnishes fertile grounds for science fiction writers, is an unachievable endeavor. And the following is for explaining that such attempts are destined to meet with dead ends.

According to Albert Einstein, traveling with very high velocities as per super-sonic contrivances, or for that matter a "Time Machine," our clocks will slow down. Meanwhile, the clocks of Mother Earth will continue to maintain their rhythm i.e., keeping their regular pace which will be quicker than that purported from the speeding T.M. Henceforth, if the time-travelers continue their phantasmic super-velocities, their "days" will be longer in comparison with the "days" on Mother Earth. Consequently, the slowing down in the times on board the T.M. will add up to hours, the hours into days and the days into years. Unfortunately, "The Slowing Clocks" anomaly was taken out of context. And here is why.

Now for the sake of understanding why the alleged time traveling to a future was theoretically tooted by many, let us accept their assumption that one year of accumulated increments generated by the lapses in time, caused by our speeding T.M., comes to equal three years of Mother Earth's time. Thus, when the time travelers return from their journey, they will be under the impression that they are arriving to a future that has happened after a seemingly one year of absence. However, this is an illusory conclusion. Simply the difference in time, is due to the skewed performance of the clocks on board the speeding T.M. Certainly both the time-travelers and the people left behind on earth have lived equal durations, i.e., three years by earthly calculations and one year by the "outer space" calculations.

Now the question is, can we consider this as traveling to a future? Undoubtedly, this phenomenon cannot be credited for such a claim.

Though not a fully concordant comparison, travelers crossing the **International Date Line**, say from a Monday to a Tuesday, cannot claim going to a future. Nor on their returning journey can they claim going to a past.

Also, for the sake of driving a point let us consider, though from a surreal point of view, the following deliberation.

The idea of going to a very far future traveling by means of ultra-velocities has its critical ardors, yet those who took upon themselves to herald the idea, stopped short from discussing the consequences of journeying with "super velocities". What they held back from mentioning, is that a machine which will take its occupants to an unknown future has to travel with velocities comparable to that of light waves. To attain such speeds, the time machine, together with its time travelers, has to go through a "**Materialistic Barrier**" to shed the bulkiness and to get transformed into quanta radiations. And presumably upon their **tele-transportation*** to "Destination Future", they should be met with **"Integrator Machines."** Hopefully these machines, if found at this other end, will reverse the process

---

* *On June 17, 2002 a TV anchorman announced the dawning of "**The Tele-Transportation Age**." In this new endeavor, Australian scientists succeeded in teletransporting on a laser beam, a mass from one place to another. Though one meter apart, the researcher asserted that longer distances will not be a problem.*

to give what is flesh to the time travelers and what is metal to the Time Machine.

Henceforth, going to a future is not going to happen, not because we cannot transform the travelers and their vehicles into fictional quanta-radiation, nor is it because our scientists haven't solved the sorting out of humans from their metallic vehicles by the supposedly available "Integrator Machines" upon their arrival, a problem that they will have to face and solve later on. But because the needed speeds are in the ambit of quanta radiations for which our bodies are unable to cope with. Nor to forbear the unexpected when the proxy quanta substituting for the time travelers, are reversibly processed by the I.Ms. This is because, unlike a homogenous piece of metal, a human body is an assortment of organs and tissues, thus permitting for mixed up turnouts instead of wholesome beings.

Surviving the future by cryogenics is another time traveling venture. It is a doubted methodology with doubted results. And for an eerie reminder, up till now, all the King's horses and all the King's men couldn't **raise the Mammoth** again. Our trials will be comparable to bringing the Cro-Magnon, the Neanderthal man, or "Lucie" to life. Unable to accommodate the futuristic changes, they will suffer from "Future Shocks" and they will be longing to go back to their old worlds, preferring them over the unfamiliar new habitats. Definitely our "Cryo-Magnons", those persons kept in deep freeze, will follow suit and will prefer their eternal frozen undisturbed slumber over the future they once longed to go to.

C. The Day Times Were Ordered to Roll Back:

The universe, according to **"The Big Bang Theory"**, is meant to expand and to go on expanding with the passage of time into futuristic happenstances. Yet, if the creator rolls back the times and orders the universe to shrink according to **"The Big Crunch Theory,"** we may be resurrected*. We will be walking out of our tombs like Lazarus and not from our mother's wombs. Henceforth, our overturned journey from the tombs to our mother's wombs will be like a videotape played in reverse.

In my opinion, rolling back the times by the Creator, allowing us to relive in a lost past, has a greater possibility to take place than to happen via a time machine claimed to be able to explore the past and

---

For a passing remark, this may explain how God will resurrect the dead on the day of reckoning by ordering "The Big Crunch". Whether this is going to happen or not, your guess will be as good as mine.

to expose a future not yet created. Going back in time to the dawn of humanity may show us how men of those early epochs revered their apparitional elderly. Knowing no better, they believed in what they were told out of respect to their old age wisdom. Unfortunately, this will destroy our blissful myths of today and may cause wars between the believers of the inherited allegories and the believers in secularism.

And while I am at it, let me borrow the mantle of Mr. Stephen King and suggest a Sci-Fi movie entitled, *"The Day Times Were Ordered to Roll Back"*. This movie will be an oeillade of bygone millennia and it will be undoubtedly the mother of all movies. Scratching, for obvious reasons, the surface of events which expanded over four thousand billion years, we will be seeing in this movie our contemporary wars in reverse order, the colonization of America, the rise and fall of empires, the resurrection of the sacrificed to Inca Gods, the advent of religions, the dismantling of the pyramids...etc.

We will be seeing what was taking place in the different geological epochs, starting with the dawning of the twenty-first century, going back to the Precambrian when the early seas were inseminated by macromolecules coming on meteorites from outer space to jump-start the early life forms. Events humanity strived, over the ages, to solve by the microscope, the telescope, and myriads of other instruments. Strived to know by examining fossil records and archaeological relics of the past and strived to understand by looking into themselves and into other beings.

In this movie, **The Book Of Genesis** will be leafed starting with the last page, going back in time to the onset of "**The Big Bang**" when nihility was ordered to split into space and time and to split into matter and antimatter. Surprisingly, or for this percept, expectedly matter and antimatter cannot exist in juxtaposition since they will cancel each other to nothingness. On the other hand, time and space are counterparts, yet if space collapses, time, its fourth dimension, will follow suit to end too in the starting point, which is nothingness. This could mean that an alleged time machine, if ever invented, will neither be able to go to auld lang syne nor to go to the future. Going through an alleged time barrier to the past will result in collapsing space and, together with it, its fourth dimension time, to the starting point of creation, i.e. nihility. And going through an alleged time barrier to the future is not going to happen because the future hasn't been conceived yet by the Creator.

Now we have to say to the movie watcher, "Please keep your seats, *The Day Times Were Ordered to Roll Back*" hasn't ended yet. The movie director, Mr. Steven Spielberg, has to close the circle of events to make a picture perfect finality. That is because we must include the other mirror image Universe; that other Universe, which was made of antimatter. In this Antithetical Universe, there are replicas of everything from antithetical galaxies to antithetical solar systems. In short, there is also an antithetical earth with antithetical geological epochs, antithetical humans, one of them will be an antithetical inventor making his antithetical time machine at the same time our inventor is making his. They will be boarding, coincidentally, their time-machines to journey into their concordant pasts to be on a collision course. **"The Big Crunch"** has been ordered to transpire.

The two universes of matter and antimatter are now racing to the staring point of creation to collide and to nullify each other to reinstate nihility "de no-vo." But let us hope that our Creator will spare us this terrible fate reminiscent to that of Sodom and Gomorrah befitting the presence of few good men among us.

Now a point of order, with my opposition to time traveling whether by time machines or cryogenics, do I want the concerned to give up their endeavors? Certainly not! In this particular field, i.e., inventiveness, everyone is entitled to his own opinion. Besides, there are many inventors who started with one thing and ended up with another useful thing. William Perkin sought of a way to make artificial quinine and ended up with aniline, an intense blue dye used extensively in industry. Charles Macintosh, working on naphtha, discovered that it liquefies rubber. Hence, he spread the liquefied rubber on cloth and invented the raincoat in 1823.

A Message in a Bottle

An old favorite with a new twist.

This bottled is tossed not into the sea, but for a first time

into a new medium, the pages of a book.

Fig. 3: Will the pages be kinder than the waves?

Hopefully, someday, "The Big Crunch" will be shown on the silvery screen.

P.S. 1: To be opened by the addressee.

P.S. 2: According to the classification of inventions presented herein, this invention is an "amendatory invention".

183

Wagdy A. Assawah

Postscript:

The Following is about a new experiment concerning the "**Big Bang**", which I found appropriate to bring to the attention of the reader.

On the 10th of July, 2001 while I was about to finish my manuscript, the news came on the TV that scientists managed to imitate the "Big Bang", the sempiternal explosion. They found that matter and antimatter originated simultaneously at the time of eruption. However, matter was produced in excess of antimatter. And when equal amounts of matter and antimatter combined to cancel each other, the portion of matter left behind, according to their apriorism, made our universe.

If that is so, with respect to our universe, then the opposite could also have happened to the antithetical universe. Meaning that the excess in antimatter left after canceling matter made the building blocks of that antithetical universe. The report concerning the Big Bang experiment doesn't change my mental image concerning the rendezvous of our time-travelers towards nihility. In fact, it reinforces the surmised.

PART III

MY MENAGERIE OF PATENTED AND NOTARIZED INVENTIONS

New opinions are always suspected and usually opposed without any other reason but because they are not already common.

John Locke

CHAPTER 5

THE RALLIED INVENTIONS

The manuscripts of inventions forwarded here-in are presented in chronological order. Each manuscript is preceded by a testimony showing its date of conception from a legalized authority. To embrace the idea of the invention, a concise recountal, highlighting its main features, is offered to help the reader to consult what interests him the most.

For truth's sake, the manuscripts are given in their original form. Though not in conformity with the edict preferred by the P.T.O., yet the disclosed idea will not escape the reader's comprehension. The only part that was purposely omitted from each manuscript was the claims. That is because the purpose here is to publish the ideas and not to patent them. Thence, the claims were found to be out of the question. Also, in few cases addenda were included to sub serve the original manuscript.

Headings of the projected inventions:

I   On how to procure spilt oil lost to sea from damaged oil tankers.

II   A simple method for desalting seawater.

III   The Bee-hive Sea-water Desalting Battery.

IV   Manually Driven Water Propulsion Device.
(A Solo-Treck, Exoskeletor, Human Performance Augmentation Device)

V   Protected Heliports to safeguard commuters.

Wagdy A. Assawah

VI(A) Evulsion Clamps to sever skin flaps.

VI(B) On how to Eviscerate a Dangling Belly Flap.

VII   A rotor dinghy for shooting rapids.

VIII   Shuttle bridges to overpass obstacles.

IX   The Manumitter

Approbatory Caveat:

Form the projected list of inventions, the reader may ask, "Your credentials show that you were specializing in Chemistry and Microbiology. How come you are treading on engineering grounds, physics domains, and wading in medical territorial waters?" This could be true for the majority of people, yet it is farthest from right concerning the perceptive and the intuitive few. Otherwise we could ask why Moses the stutterer was chosen to speak to God. Or why

Mohammed the orphan, was chosen to be the messenger of the "The Supreme Creator".

To say that I studied hosts of chemistry courses and I studied myriads of biological subjects from viruses to man before I got my degrees from Cambridge University and Nottingham University, U.K. is not a justification. And to say that I had a computer memory storing whatever I have read in different endeavors over the 52 years after my graduation is a sullen answer.

For undoubted examples of people crossing such boundaries, I shall ask you to recall the earlier mentioned story of Thomas Edison. And to go back to that of Alfred Wegener, who published his continental drift dogma though he was a meteorologist, not a geologist. Their stories show that revelations know no boundaries.

Also for an educated justification, erudites and the sage predicate, "Convince us that your acquired cognition was taken to heart and that you understand the ins and outs of your suggested premonition. This in itself will out weigh by far the bestowed certificates decorating the walls of those who sat on the glory of their degrees and who took refuge in their diplomas, thinking that they were the cessation and not the initiation".

Last but not least, I'd like to give a hint at the importance of simplicity and to show that it is the best policy in the business of inventiveness. Hence, I find it worthwhile to point out that a streak of simplicities is shared by a number of my inventions, as evidenced by the following:

- The leaning V-shaped tubular lenses used for desalting sea-water.

- The coupling of propellers in juxtaposition to the heels of users thrusting them by either a hand-driven or a motorized-driven water propulsion device.

- The explicit techniques targeting the evisceration of skin flaps from human bodies by the invented Snare-Tackles.

Such simplified methodologies may look for a first countenance unconvincing to those unfamiliar with the art, yet they will be implicit to the versed who intuitively envisage that things that may look subtle in the beginning will be significant in the end.

I. HOW TO PROCURE SPILT OIL LOST TO SEA FROM DAMAGED OIL-TANKERS

P.T.O. Disclosure   Document No., 296088

Mail Room Date   November 18, 1991

A Concise Recountal:

Reacting to mishaps caused by oil lost to the sea, a way was thought of to recover this oil and at the same time minimize its harmful effects on the environment. Apart from being disastrous to the fauna and flora of the shoreline, it is also calamitous to the sea dwellers, from phytoplanktons to kelps and from zooplanktons to whales.

In essence, the invention suggests rounding up the lost oil in big puddles with the help of vertically held curtains. The oil together with some seawater is then scooped by means of "skimmer tugs". The mixture is pumped into transparent separators. The oil in the separator is then decanted and collected and the seawater is returned to the sea. This method is most suitable for recovering oil in sheltered ports and may not work well in open seas, especially if the weather is not cooperating.

PATENT APPLICATION

OF

WAGDY A. ASSAWAH

FOR

ON HOW TO PROCURE SPILT OIL LOST TO SEA FROM

DAMAGED OIL-TANKERS

Oil slicks from broken oil tankers coming in or going out of harbors is a mishap that can take place anywhere any time. They are deadly evils and they waste a desperately needed energy source. For this reason, the author is forwarding in this dissertation an illustrated method aiming at skimming spilt oil together with some seawater. The oil, with the help of sizable transparent "separators," is then decanted and collected while the seawater is returned back to the sea.

It is hoped that this method will, not only be beneficial to the environment, but it will also be prolific to the oil companies. As for the environment, skimming the oil will spare to a great extent the harmful effects to planktons and to microhabitats, and it will prevent most of the agonizing deaths to the sea dwelling creatures. As for the oil companies, procuring spilt oil will reduce losses in the valuable cargo and will minimize drastically the cleaning up expenses.

The Technique For Procuring Spilt Oil:

The technique adopted for procuring spilt oil lost to the sea, can be better comprehended with the help of appropriate drawings. The following four figures, together with the given legends, are for explaining the different steps to achieve this purpose. The portrayed figures comprise:

- Figure (1) for displaying a side view of the skimmer tug. It demonstrates how the components of the scooping mechanism and the "separator" are mounted on the front part of the tug and how the procured oil is collected and delivered to a towed barge.

- Figure (2) for showing the details of the transparent blimp-shaped "separator" and for explaining its role in procuring oil lost to the sea.

- Figure (3) for displaying the top view of the keel of the skimmer tug, to indicate the proportions and the locations of the scoops, the pumps, and the system of pulleys and levers to one another and in relation to the "separator".

- Figure (4) for demonstrating how to round up oil puddles with the help of appropriate floats. It is also for exhibiting how the skimmer tugs are wedged in special openings in the encircling float to scoop the besieged oil.

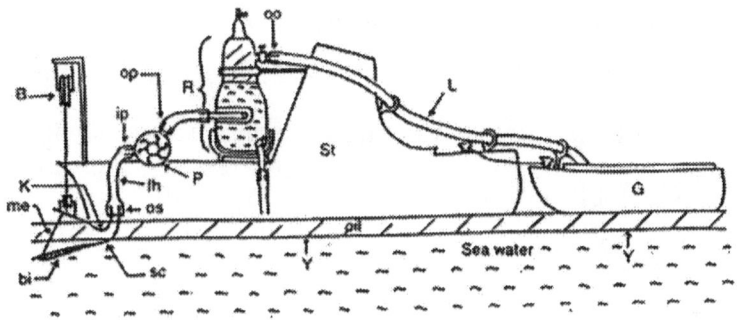

Figure (1) Displaying a side view of a skimmer tug. It is also showing the scooping mechanism (sc), (P) and (B) and the transparent "separator" (R) responsible for procuring oil lost to the sea. The regained oil is then delivered via a long hose (L) to a towed barge (G).

Legend For Figure (1):

sc   Side view of the longitudinal biflanked scoop (see figure 3). It has a wide mouth opening embracing the oil/seawater diphase. The oil/seawater mixture is sucked by the pump (P) via a flexible hose (fh) connecting the outlet of the scoop (os) with the inlet (ip) of the pump (P).

195

me   An oblique mesh fitted in the opening of the wide mouth of the scoop to prevent dead fish and other floating objects from going into the system, hence obstructing the flow of the sucked oil/seawater mixture.

bi   Built-in weights on the underside of the lower lip of the scoop to stabilize it and to keep it embracing the oil/seawater diphase especially when the tug is at work.

K   Side view of the lengthwise kink at the neck of the scoop to prevent, as much as possible, air from going into the system together with the skimmed oil/seawater mixture.

P   Suction/compression pump. Each skimmer tug has two pumps, one on each side of its keel. The pumps are responsible for both sucking the oil/seawater mixture and injecting it through their outlets (op) into the blimp-shaped transparent "separator" (R).

B   The system of pulleys fixed to an overhead spanning metal beam. The pulleys are for adjusting the position of the scoop in relation to the oil/seawater interface (Y). With the help of the pulleys and the levers operating them, the scoop can be hoisted in the air for the skimmer tug to move at ease from one place to the other.

L   The long hose originating from the oil outlet (oo) of the separator (R) delivering the procured oil to the towed barge (G).

R   The blimp-shaped transparent "separator", the detailed structure of which and the role played by it, are the subject matter of figure (2).

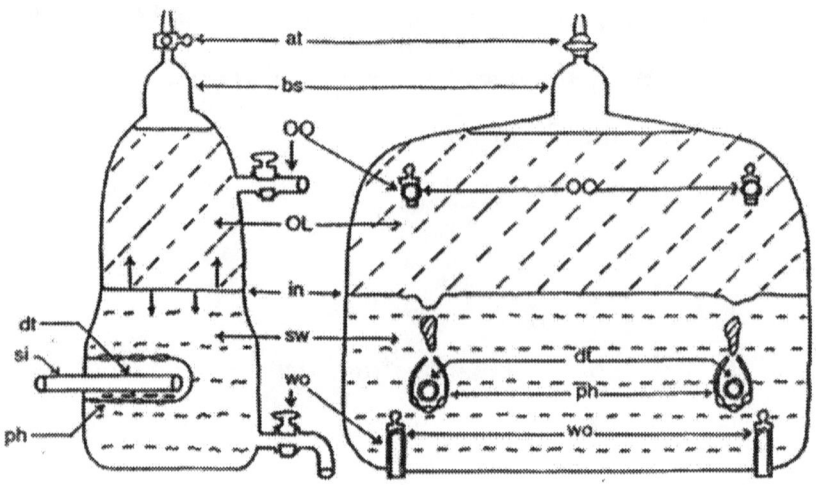

Figure (2) Showing the detailed structure of the blimp-shaped separator.

Side View of Separator (R) Displaying the positions of the delivery tube (dt) together with its perforated hood (ph), the oil outlet (oo) and the sea-water outlet (wo), both fitted with appropriate taps.

Face View of Separator (R) On the nearer side of which there are two oil outlets (oo) and two sea-water outlets (wo), while on the far side there are two delivery tubes (dt) enveloped by two perforated hoods (ph).

Legend For Figure (2):

R.   The blimp-shaped separator. It takes in the oil/seawater through two inlets (si). There, the mixture separates into two distinctive layers, an upper oil layer (OL) and a lower seawater layer (sw). This allows for easy decantation of the oil layer.

dt.   The delivery tube, a continuation of the separator inlet (si), injecting the sucked oil/seawater mixture into the separator (R).

ph.   A perforated hood covering the mouth part of the delivery tube (dt). It is for lessening the turbulences resulting from injecting the oil/seawater under pressure, thus allowing for a gentler diphase separation.

bs.   A bell-shaped structure for collecting air bubbles. The kink (K) in the scoop may fail to exclude all the air accompanying the oil/seawater mixture. The bell-shaped structure is provided with an apical tap (at) for releasing the collected air.

Wagdy A. Assawah

in.   The interface of the oil/seawater diphase. It is expected to fluctuate, as shown by the arrows. It should be maintained within certain limits by carefully manipulating the streams of procured oil discharged through the outlets (oo) and the streams of spent seawater through the outlets (wo).

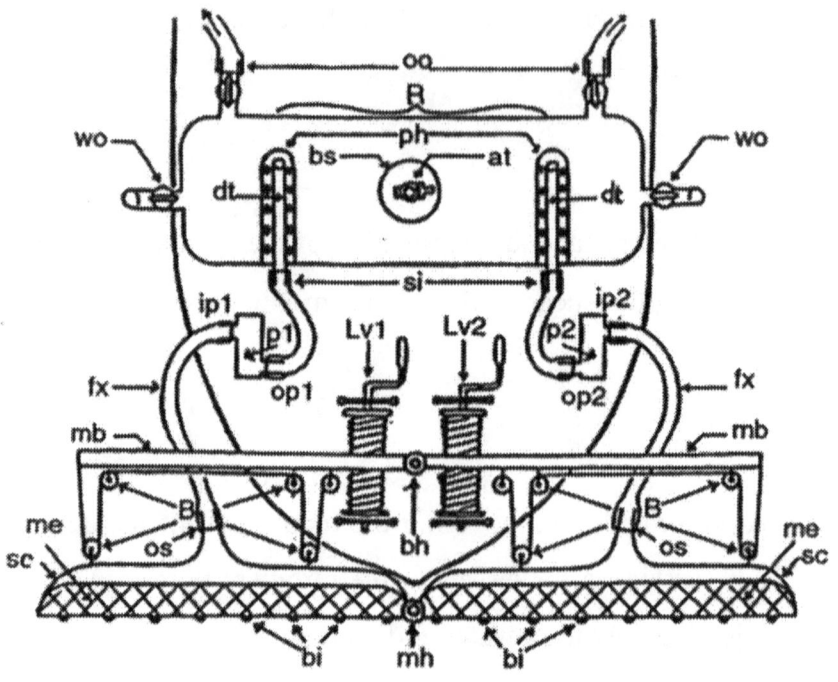

Figure (3) Top view of the keel of a skimmer tug showing the positions of the blimp-shaped separator (R), the double mouthed scoop (sc) and the two suction compression pumps P1 & P2.

Legend For Figure (3):

sc   The wide mouthed scoop of two flanks held together by a medial hinge (mh). The hinge allows the scoop to take a more or less v-shaped configuration if maneuverability for oil scooping justifies that. The mouth of the scoop is covered by a mesh (me) and it is kept dipping and embracing the oil/ seawater diphase with the help of built-in weights (bi).

B   The system of pulleys fixed to an overhead spanning metal beam (mb) of two arms held together by a medial hinge (bh).

LV1,   and LV2 are two levers fixed to the deck of skimmer tug. Both systems of pulleys and levers are used to adjust the position of the scoops to achieve maximum skimming capability. They can also keep the scoops up in air for the skimmer tug to cruise in open sea with ease from one place to the other.

bs   Top view of the bell-shaped structure, that is needed, for trapping air that may lurk into the system. The trapped air, thus collected, can be released through the apical tap (at).

P1,   and P2 are two suction/compression pumps, one on each side of the keel. The sucked mixture passes from the outlets of the scoop (os) to the inlets of the pumps (ip1) and (ip2) via flexible hoses (fx). From there the oil/seawater mixture is compressed through the outlets of the pumps (op1) and (op2) to the inlets of the separator (si). These inlets extend into the latter in the form of long delivery tubes (dt).

oo.   The oil outlet originating from the upper part of the separator. It is for discharging the procured oil via a delivery hose (L) to a tow barge (G) as shown in figure (1).

wo.   The sea-water outlet originating from the lower part of the separator is for spending unwanted seawater to the sea. Both the oil outlets (oo) and the seawater outlets (wo) are provided with taps to regulate the outgoing streams of oil and seawater respectively.

R   Top view of the separator showing the two delivery tubes (dt). Each is enveloped by a perforated hood (ph) to slow down frothing caused by the compressed incoming oil/ seawater mixture. In the separator (R), the oil/seawater diphase separates into an upper lighter oil layer and a lower heavier seawater layer. The oil layer is then decanted and delivered via the outlets (oo) to the tow barge (G), while the dispensable seawater is spent via the outlets (wo) to the sea.

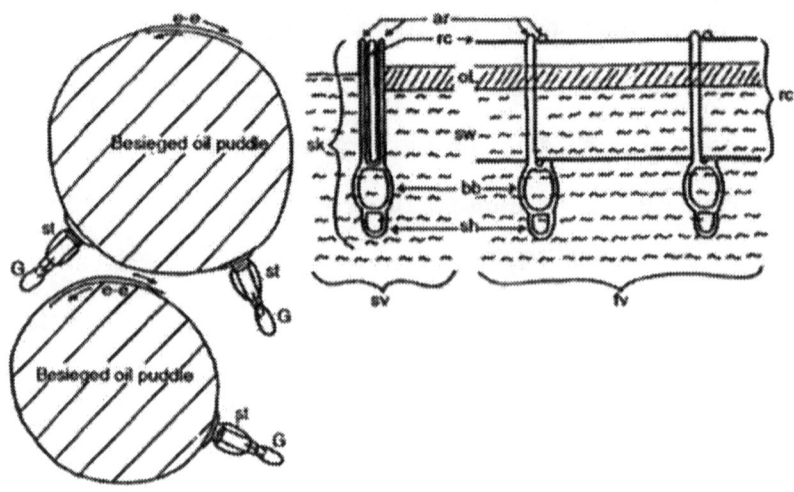

Figure (4b) Showing diagrammatic representations of two oil puddles rounded up by circular floats with skimmer tugs (st) at the periphery scooping the split oil. An oil puddle could be a mile in diameter.

Figure (4a) Showing the components of the float and its position in relation to the oil layer and the sea-water underneath it. FV-Face view of part of the float displaying the two components, the sinkers (sk) and the curtains (rc), viewed from inside the oil puddle. SV-Side view of one of the sinkers (sk) in an upright position holding the vertical rubber-curtain (rc) between its two arms (ar).

Legend For Figure (4)A:

rc   The thick rubber curtain fitting between the arms of the sinker (sk). The upper edge is sticking out in the air above the oil surface and the lower edge is dipping in the seawater well beyond the common surface of oil/seawater interface.

ar   The two parallel arms of the sinker (sk) holding the curtain (rc) in an upright position.

bb   A bubble filled with air (or helium) to keep the sinker (sk) at a certain depth.

sh   The sinker's head enveloping a calculated weight to keep the sinker in a standing position.

oL   The oil layer of the besieged oil puddle.

sw   The seawater beneath the oil layer.

Wagdy A. Assawah

Legend For Figure (4)B:

The figure is showing the skimmer tugs (st) wedged at certain openings in the upright curtains of the float. The openings are designed to accommodate the scoops to allow for sucking the oil/seawater mixture from the oil puddle. The procured oil from the earlier described separator, is then delivered to the towed barges (G). As the besieged oil gets less and less due to the skimming process, the float is constricted to lessen the area of the oil puddle. This is done by pulling the overlapping free-ends (e-e) in the direction of the arrows shown in the diagram. This lessening of the oil puddle area, will keep the oil layer thick enough so as to be suitable for scooping. Procuring of the oil and constricting of the float are to be maintained till all the spilt oil is regained from the sea.

II. A SIMPLE METHOD FOR DESALTING SEAWATER

P.T.O. Filing Date 11/25/91   Serial No., 07/765,202

Foreign Filing License   Granted 02/06/1992

A Concise Recountal:

Knowing that the world is plagued with repeated droughts and shortages in fresh water, as happening in Middle Eastern Countries and in Spain, a method was thought of to desalt seawater affordably to provide potable water for drinking. Using several of the presented here-in environmentally friendly apparatuses around the clock, should also help in growing crops, if not in fields, at least in hydroponics.

In essence, the suggested invention comprises a rectangular metal basin holding shallow seawater that is continuously, yet slowly replenishable to prevent salting out. Desalting the seawater is then affected by the use of V-shaped leaning tubular lenses and heat inductors annexed to the rectangular metal basin. The V-shaped leaning tubular lenses with cooler seawater percolating in them are suspended in inclined positions at pre-determined heights over the rectangular basin to downcast lengthwise foci leveling at the surface of the seawater contained in the metal basin. The formed hot foci raise the temperature of the seawater in the basin. Also the annexed heat inductors help in heating up the contained water in basin by creating inductive currents. Due to the effect of both the V-shaped leaning tubular lenses and the heat inductors, evaporation is enhanced and is followed by condensation on the cooler underside of the V-shaped leaning tubular lenses. Responding to gravity, the formed water films run towards the lower ends of the lenses where they are helped with protuberances to form droplets of water and to shed them in a receiving trough.

This seawater desalting battery does not need an outer source of energy to accomplish its goals. It uses sun rays to heat the seawater in the basin. It also uses a windmill driven pump to replenish the leaning tubular lenses with cooler seawater. Hence there is no by-products that may pollute the environment or contribute to the green house effect.

PATENT APPLICATION

OF

WAGDY A. ASSAWAH

FOR

A SIMPLE METHOD FOR DESALTING SEAWATER

Severe shortage in water for drinking and irrigation, especially in arid zones, warrants the exploitation of any means to justify easing the situation. No wonder then, desalting seawater, is and will continue to be a dream of humanity especially in a thirsty world plagued with repeated droughts. And, though we are in the age of microchips, gene slicing, cyclotrons and the likes, the author is forwarding a simple battery for desalting seawater. This battery may prove indispensable for the poorer communities, which cannot afford expensive, elaborate, and problematic desalination plants.

Wagdy A. Assawah

THE SEAWATER DESALTING BATTERY:

(Figure 1):

The parts of this battery comprise:

The Leaning Tubular Lenses:

Each leaning tubular lens is a long hollow tube (T) with a bulbous base on one end and a wider part at the other end. The tube is provided with a peg (h) emerging from the upper side. The peg is utilized for suspending the Tubular lens in a special manner. At the bulbous end, there is an inlet (n1) through which seawater is introduced. At the wider end, there is an outlet (n2) through which a stream of seawater is allowed to seep to the ground on which the battery is installed. A horizontal tube (H), seen in figure (1) in transverse section, is overriding and connecting with the inlets (n1) of the different tubular lenses. It is used to replenish the tubular lenses with the cooler sea-water.

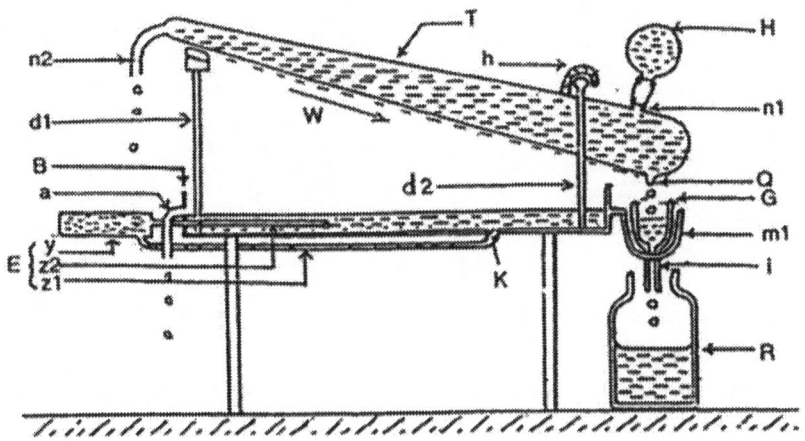

Figure (1) Side view of the sea-water desalting battery.

T.   The tubular lens, showing two openings (n1) and (n2), a peg
      (h), and a protuberance (q).

B.   Metal basin holding the sea-water. It has a spout (a), a front
      stand (d1), and a rear stand (d2).

E.   The heat inductor, showing the black boxlike structure (y)
      and the long limb (z1) and the short limb (z2).

G.   The trough with outlet (i) and supported with holder (m1).

R.   The reservoir.

W.   An arrow indicating the direction of the condensation.

H.   The horizontal tube in T.S., connecting with inlet (n1).

The Metal Basin:

In the metal basin (B), seawater is kept shallow and slowly flowing at all times. The basin is lifted above the ground by means of appropriate legs to keep it leveling at a suitable height. To maintain the contained seawater at a certain shallow depth, the basin is provided with a spout (a) for getting rid of excessive seawater. The basin is also equipped with two adjustable front and rear stands (d1) and (d2). The front stand (d1) has a horizontal padded part for accommodating the wider ends of the tubular lenses. The rear stand (d2) has a horizontal part on which the pegs (h) hook on. The stands are for keeping the leaning tubular lenses in an inclined position and at a predetermined height from the basin.

The Heat Inductors:

A simple heat inductor (E) is made of a hollow black boxlike structure (y), having two tubular limbs (z1) and (z2) placed one above the other. The limb (z1) originates from the bottom of the basin near its rear end with an orifice (k) through which seawater sinks. The other end of (z1) opens on the lower side of the boxlike structure of the heat inductor. The limb (z2) is shorter than limb (z1), originates at the upper side of the boxlike structure, then it passes through the front wall of the basin and opens beneath the surface of the seawater contained in the basin.

The Trough And The Reservoir:

For supporting the trough (G), the basin has two curved metal holders (m1) and (m2) to keep the trough suspended underneath the protuberances (q). The protuberances help the water droplets to form and to drop into the trough. The latter has an outlet (i) through which the collected water seeps to an underlying reservoir (R).

The Making Of The Leaning Tubular Lens:
(Figure 2):

Each leaning tubular lens is intended to collect the parallel sun rays falling on its upper side to produce a lengthwise focus facing the underside of the leaning tubular lens. Though inclined, the leaning tubular lens is shaped in such a way so as to form a straight longitudinal focus leveling at the surface of the seawater contained in the metal basin.

The gradual changes in the shape of the leaning tubular lens can be comprehended by making several transverse "cat-scan" sections (Figure 2) at certain plaines, namely 1, 2, 3, and 4. The drawings in (F) correspond to the cat-scans at these plaines. They show that both the shape of the tube and the curvature at a particular site, changes from one plaine to the other. The displayed changes in curvature indicate that the more the convexity of a plaine, the shorter its focus becomes, and the less the curvature of a plaine, the longer its focus becomes. In other words, the focal length at a particular plaine is inversely proportional with the convexity of that particular plaine. Henceforth, these changes affect the production of longer and longer foci as we go from the bulbous end to the wider end of the tube. The produced linear focus is meant to coincide and to level at the surface of the seawater contained in the basin in spite of the inclination of the tubular lens.

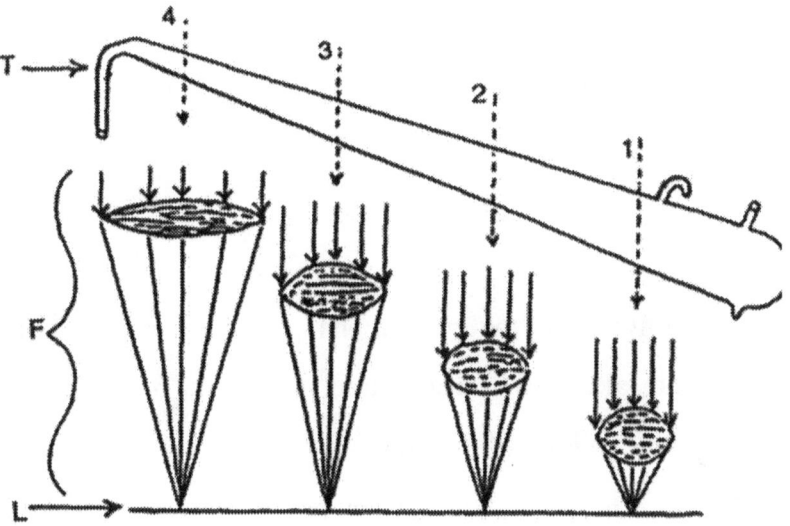

Figure (2) Showing the gradual changes in shape of a tubular lens.

Wagdy A. Assawah

T.   A tubular lens in an inclined position. The numbers 1, 2, 3, and 4 show the plaines at which transverse section are made.

F.   The corresponding transverse section at the previously indicated plaines.

L.   The lengthwise focus paralleling at the surface of the seawater contained in the basin.

To figure out the gradual changes in a leaning tubular lens, one should take into consideration, the angle of inclination and, the refractive index of the material of the tube and its length, a job made easy with the help of a computer. Thus manufacturers can make bigger tubes for bigger batteries and smaller tubes for smaller batteries.

Operating The Battery:

(Figure 3)

The inlets (n1) of the different tubular lenses of a battery are fed by the comparatively cooler seawater from the horizontal tube (H) overriding the inlets. Supplying seawater is carried out by means of a windmill driven pump (P) which retrieves the seawater from a shallow well (s) constructed by the seashore. The group of tubular lenses, with the sun rays falling on their upper sides, will form the linear foci (L) extending from the front end of the basin to its rear end. The relatively hot foci, heat up the seawater in the basin and evaporation is enhanced.

To create considerable difference in the temperatures of both the seawater in the basin and the tubular lenses, the basin is equipped with several heat inductors (E), which like the leaning tubular lenses are also kept exposed to the direct sunlight. The structure of the heat inductor allows the cooler seawater with the relatively heavier density to sink into the orifice (k) of the limbs (z1) and to be carried to the black boxlike structure (y) where it heats up. The seawater, now hotter, becomes relatively lighter in density and forms an upper layer inside the boxlike structure. From there, it flows out through the limb (z2) and gets out of its orifice which opens beneath the surface

Wagdy A. Assawah

of the seawater contained in the basin. In this way, an induction current is created. So, each time the seawater is recycled inside the heat inductor, it gains more and more heat. The pumped heat into this system, which is done by the different heat inductors, provides for the necessary latent heat of evaporation.

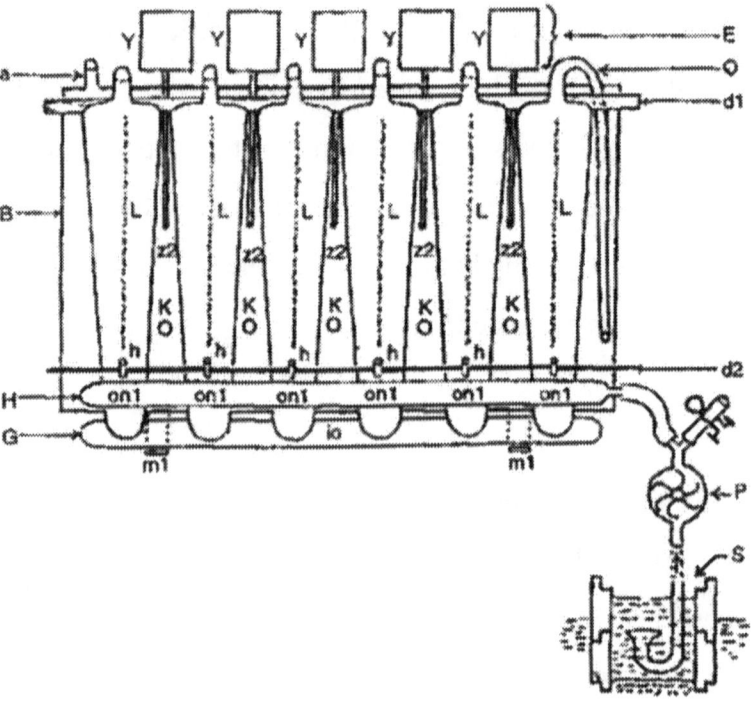

Figure (3) Top view of the desalting sea-water battery.

B.   The metal basin showing spout (a), the front stand (d1) and the rear stand (d2).

E.   The heat inductor showing the black boxlike structure (y), limb (z2) protruding into the basin and (k) the orifice of (z1).

L.   The lines representing the lengthwise foci when the sun is in the zenith.

H.   The horizontal tube overriding the inlets (n1).

G.   The trough showing an outlet (i) and the two metal holders (m1).

P.   The windmill driven pump, retrieving seawater from well (s).

Wagdy A. Assawah

O.   The rubber tubing stooped into the basin to compensate for
the loss in sea-water.

Due to the effects of both the relatively hot foci and the induction currents made with the help of the heat inductors, water vapors keep continuously forming. The vapors then rise, to meet with the relatively cooler undersides of the tubular lenses where they condense. The formed water films respond to gravity and glide towards the bulbous bases of the leaning tubular lenses. The protuberances (q) projecting out on the undersides of the bulbous bases, help the water droplets to form and direct them to fall into the trough (G) henceforth to the reservoir (R). In fact, one may look upon the seawater desalting battery as a microcosm in which we are making our own clouds and milking them.

At all times the sea-water in the basin should be free running and should not be allowed to get concentrated, to avoid salting out. When the sea-water in the basin gets low due to evaporation, sea-water from one of the outlets (n2) of a tubular lens is allowed, with the help of a long rubber tubing (O) to compensate for such loss. Obviously, the rubber tubing should not be left continuously inside the basin. Adding excessive amounts of seawater from the tubular lens, lowers the temperature in the basin, and this in its turn slows down the evaporation, thus affecting the yield.

The Foundations Targeted In The Making Of This Battery:

It is noteworthy to point out that the established theories of physics, put to use in planning this sea-water desalting battery, include:

1.   Convex lenses, or for that matter, leaning tubular lenses to collect sun rays in a burning hot focus.

2.   Evaporation is enhanced when a liquid (in this case water) is heated.

3.   Water vapor condenses upon meeting cooler surfaces.

4.   Water films run on sloping surfaces, responding to gravity.

5.   The protuberances provide the needed surface tension for water droplets to form and to fall in a designated place.

6.   The more the difference in temperature between a relatively hot body of water and a relatively cooler surface, the more the yield in condensation.

7.   The slight differences in the density of a liquid caused by heat, result in developing currents of induction. The currents move about to homogenize the acquired energy in the heated liquid.

P.S.   The above mentioned leaning tubular lens, is an invention in its own right, entitling the inventor to name it after his name, i.e., **"Assawah Lens"**.

In a descriptive statement from the "Bureau Of Reclamation" as shown by the included here-in "Addendum A": "Using a tubular lens with a varying cross section is an ingenious idea to concentrate the sun's energy".

III. THE BEEHIVE SEAWATER DESALTING BATTERY

P.T.O. Disclosure   Document No., 293614

Mail Room Date   October 21, 1991

A Concise Recountal:

Earlier, the author presented a simple battery for desalting seawater. This was achieved by using leaning tubular lenses and heat inductors fitted to a metal basin containing the seawater. This simple battery is apt to yield unsatisfactory amounts of desalted water. To obtain a generous yield, more tubular lenses per battery were to be affixed to the battery and several batteries were to be linked together. The hexagonal shape proved to be convenient for this purpose, and the resulting composite configuration was reminiscent of a bee-hive. Hence the name "bee-hive" was coined to distinguish this battery from its forerunner.

Added to getting a generous yield through making a composite battery, one can also manipulate the batteries to give a continuous yield around the clock. For this purpose, the heat inductors can be fitted with electrically operated immersion heaters, so the batteries, whether simple or of the beehive type can perform around the clock.

Wagdy A. Assawah

PATENT APPLICATION

OF

WAGDY A. ASSAWAH

FOR

THE BEE-HIVE SEAWATER DESALTING BATTERY

Introduction:

In an earlier communication, the author discussed how to desalt seawater by means of a simple desalting battery. It comprised leaning tubular lenses and heat inductors and it was operated by solar energy and a windmill driven pump. Obtaining desalted water was brought about by enhancing water evaporation in the battery, then collecting the condensate.

A single desalting battery, under the best conditions of sunshine and cooler seawater, yields little quantities of desalted water. So, it was then logic to expect that greater yield could be obtained by increasing the number of tubular lenses in a battery and by connecting several batteries together. To accomplish this, the circular or the hexagonal shapes offer promising possibilities. Such shapes allow for grouping individual batteries into a composite multi-battery reminiscent of a bee-hive.

Building the bee-hive battery:

The bee-hive pattern thus thought of, involved linking six individual batteries into one composite multi-battery. Not only did this necessitated changes in the displacement of the different components of the individual battery, but also it necessitated changes within the composite multi-battery.

Wagdy A. Assawah

The changes involved in the individual battery: (Figure-1)

The Basin:

Unlike the rectangular basin in the simple battery, the basin is here hexagonal with a hollow central part delimited by an inner rounded wall (rw). Concerning the front and rear supports (d1) and (d2), instead of running from one side to the other in the simple battery, they are here conforming with the hexagonal contour. The support (d1) parallels with the outer walls of the basin, while the support (d2) forms a hexagon encircling the round wall delimiting the central hollow part of the basin (Figure 2).

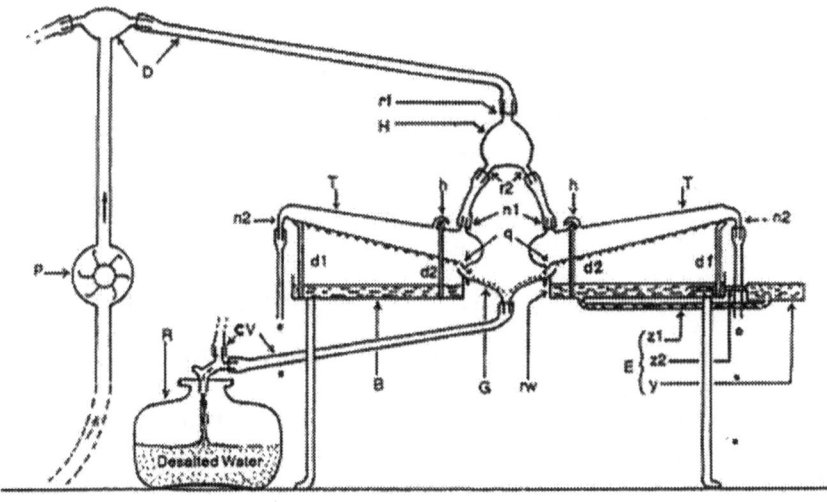

Figure 1 Side view of one of the units making up the composite beehive battery.

T. The tubular lenses held inclined by the supports (d1) and (d2). Also showing pegs (h), protuberances (q), inlets (n1) and outlets (n2)

B. The metal basin with the contained sea water.

P. The pump and the system of diverging tubes (D).

G. A funnel shaped trough supported by the rounded wall (rw).

E. A heat inductor and components (y), (z1) and (z2).

H. A pear-shaped distributor showing Inlet (r1) and Outlets (r2).

R. The common reservoir and the system of converging tubes (CV).

227

The Leaning Tubular Lenses:

Unlike the side-by-side linear arrangement of the leaning tubular lenses in the rectangular simple battery, the leaning tubular lenses are here arranged in a radiating pattern around a common center. This allows for lesser gaps between the tubular lenses and for effective interception of the rising water vapors. The leaning tubular lenses, whether they are components of the simple battery or they are components of the hexagonal battery, are supported at a predetermined height in an inclined position. The wider ends of the tubular lenses lean on the horizontal padded part of the front support (d1), while their narrower ends are held in position with the help of the pegs (h) hooking on the horizontal part of the rear support (d2).

The Seawater Feeding Mechanism:

The feeding, in the simple battery, was accomplished by a horizontal tube overriding the inlets (n1) of the leaning tubular lenses. Meanwhile, the feeding in the current individual battery is accomplished by means of a pear-shaped structure (H). This structure receives the seawater from a windmill driven pump (P) through an apical opening (r1). It has also several outlets (r2) corresponding in number with the numbers of the tubular lenses. The outlets (r2) connect with and supply the seawater to the underlying tubular lenses through their inlets (n1). Excessive seawater in the tubular lenses, is then allowed to flow out through the outlets (n2) located at their wider ends and hence to the ground on which the bee-hive battery is mounted.

The Trough:

Collecting the desalted water resulting from the condensation, was carried out in the simple battery by means of a longitudinal trough suspended underneath the bulbous bases of the leaning tubular lenses. To achieve this aim in the bee-hive battery, each individual battery is provided with a funnel-shaped trough (G). Here the rounded rim of the trough is supported by the rounded wall (rw) delimiting the central hollow part of the basin. The rim of the trough is designed to encompass the bulbous bases of the leaning tubular lenses. Thus, the trough is in a position to receive the desalted water droplets helped to form by the protuberances (q) of the leaning tubular lenses. The received water eventually seeps via the stem of

Wagdy A. Assawah

the funnel-shaped trough and is delivered via a rubber tubing to the common reservoir (R).

The Heat Conductors:

The simple battery is equipped with heat inductors (E) to bring about induction currents. These enrich the seawater in the basin with the heat needed to enhance evaporation. The currents of induction are products of the special arrangement of the limbs $(z1)$ and $(z2)$. The longer limbs $(z1)$ obtain their supply of seawater through the orifices (K) from the cooler parts of the battery. The seawater is then carried to the black boxlike structure (Y) where it heats up. Thus heated, the seawater then flows out through the orifices of the shorter limbs $(z2)$ which open just underneath the surface of the seawater contained in the basin.

In the bee-hive battery, the heat inductors (Y) are arranged on the three outermost walls, namely A, B and C of each individual battery (Figure 2). To transport the seawater from the cooler parts of the battery, some of the limbs $(z1)$ are skewed in such a way so as their orifices (K) open around the wall (rw), delimiting the central hollow part of the basin.

The Changes Involved In the Composite Battery: (Figure 2)

The Pump:

The linking of individual batteries made it possible to use one pump (P) to serve the different units of the bee-hive battery. The pump (P) retrieves the seawater and dispenses it through a system of six diverging rubber tubings (D), one for each pear-shaped distributor (H). This latter, in its turn replenishes the tubular lenses with the continuously running cooler seawater.

Wagdy A. Assawah

The Reservoir:

The linking of the individual batteries made it also possible to use one common reservoir (R) to work for the different units of the bee-hive battery. The reservoir (R) receives the desalted water seeping from the stems of the funnel-shaped troughs (G) via a system of six converging rubber tubings (CV).

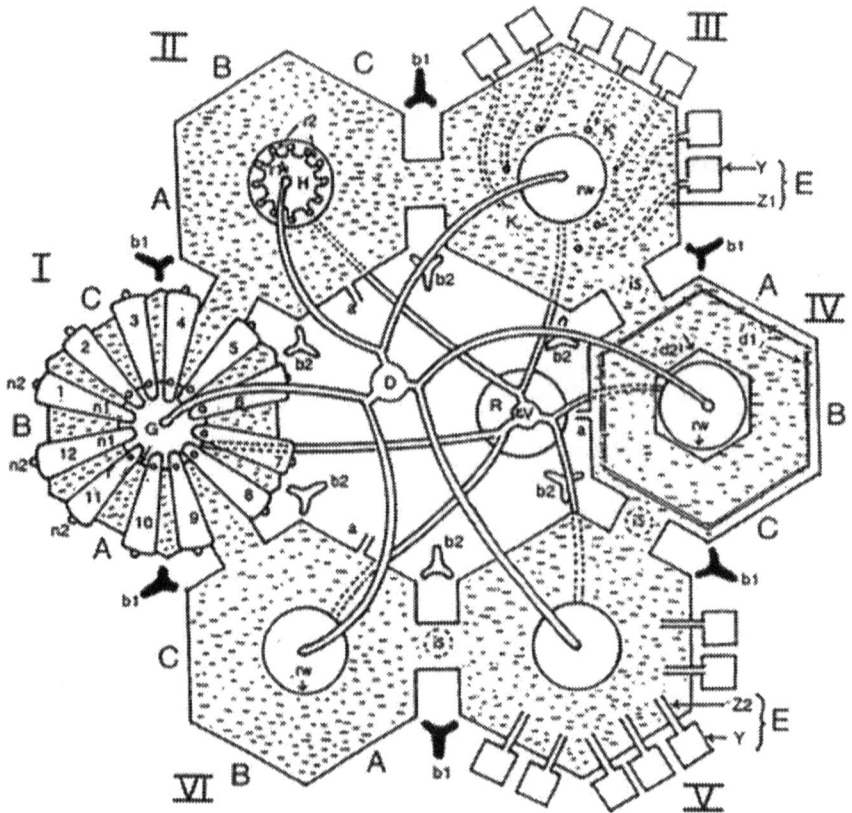

Figure (2) Top view of the Bee-hive Battery
See the following page for legend

232

Other Minor Changes:

These include, connecting the individual batteries with each other by isthmii (is). As for the spouts used for getting rid of the excessive seawater in the basin, only three are made for the six connected batteries. The spouts (a) are installed on the inner most walls, one on every other hexagonal basin.

Legend For Figure (2):

N.B. The units I to VI do not show the complete structure of any individual battery. For the sake of simplicity, each unit displays one structure or the other. Consequently, a lengthy legend was inevitable.

I   Showing the radiating arrangement of the twelve leaning tubular lenses with their narrow ends encompassed by the underlying trough (G).

II   Showing the pear-shaped distributor (H), the top inlet (r1), and the twelve outlets (r2).

III   Showing the arrangement of the heat inductors (E) on the outer most free walls of the basin designated by A, B and C. Unit III shows also the courses of the limbs (z1) and that their orifices (K) open in the vicinity of the rounded wall (rw).

IV   Showing the horizontal parts of support (d1) paralleling with the outer wall of the hexagonal unit and the horizontal parts of the support (d2) forming a hexagon encircling the inner rounded wall (rw) of the basin.

V   Showing the heat inductors (E) and their shorter limbs (z2).

VI   Showing three sides on which the heat inductors are positioned, two sides connecting with adjacent units by isthmii (is) and an inner most side displaying a spout (a).

Other structures found outside the individual units that operate the composite multi-battery include:

D.   The central part of six diverging rubber tubings comprising the system responsible for carrying the seawater coming from the pump (P) and handing it over to the pear-shaped distributors (H).

CV.   The end part of six converging rubber tubings comprising the system responsible for collecting the desalted water from the funnel-shaped troughs (G) and handing it over to the common reservoir (R).
b1 and b2 representing the outer and inner legs respectively. They support and level the bee-hive battery at a suitable height from the ground.

Wagdy A. Assawah

Operating The Bee-hive Battery:

The roles played by both the leaning tubular lenses and the heat inductors in enhancing seawater to evaporate, to condense, and hence to collect the condensate, were previously explained in the earlier communications entitled, "A simple method for desalting seawater". Operating the bee-hive battery is no different from operating the previously mentioned simple battery. Nevertheless, emphasis on the following should be singled out.

1. Running of the cooler seawater through the tubular lenses should be maintained at all times.

2. When the seawater in the connected basins gets low due to evaporation, it is then allowed to be compensated for. This is carried out by stooping one of the rubber tubings from one of the outlets (n2) into one of the basins so that it replenishes the basins. The flow of seawater from the stooped rubber tubing can be regulated by a screw-clamp, so that the incoming seawater in the system compensates for the amounts lost through the spouts (a) and lost to evaporation.

236

3.   The batteries for desalting seawater, whether in the simple or the bee-hive batteries, are expected to work during the daytime with a peak in the middle of the day. This shouldn't be the case, and the presence of the sun shouldn't be a limiting factor. In places where electricity abounds, the heat inductors can be manufactured with built-in immersion heaters, so that the desalting batteries can be worked around the clock.

Wagdy A. Assawah

Addendum (A):

A letter to the inventor from the "Bureau of Reclamation" discussing "The Simple" and "The Bee-hive" seawater desalinator batteries.

 United States Department of the Interior

BUREAU OF RECLAMATION
Denver Office
P.O. Box 25007
Building 67, Denver Federal Center
Denver, Colorado 80225-0007

IN REPLY REFER TO:

D-3710

Dr. M. Wagdy Assawah
545 Wistar Road Apt. H22   MAY 29 1992
Fairless Hills PA 19030

Subject: Apparatus for Desalting Sea Water (reference your letter of
          March 3, 1992) (Desalting) .

Dear Dr. Assawah:

Your letter dated March 3, 1992, to the Department of Agriculture requesting a
review of documents relating to an apparatus for desalting sea water has been
referred to this office. Our staff has reviewed the documents and have the
following comments:

Technical Feasibility

 • The apparatus uses the same principles of thermodynamics that are used by
   existing solar stills as shown in the enclosed report, so it is agreed
   that the concept is technically feasible.

 • Using a tubular lens with a varying cross section is an ingenious idea to
   concentrate the sun's energy. However, as in any solar collector system,
   collector panels that slope to the north collect very little heat unless
   installed at the equator. This would not be a problem in the proposed
   rectangular arrangement if it is constructed to face the south, but in
   the bee-hive arrangement, some of the panels that would have to face the
   north would collect very little heat.

 • If the same total amount of energy is going into the same depth of water
   in the basin as in a conventional solar still, very little is gained by
   concentrating the heat because of the large thermal capacitance of the
   water. However, if the depth of the water is kept at a shallower depth
   at the focal point to allow higher localized temperatures, it can be seen
   that this type of solar still should have a higher efficiency than other
   stills.

   Owing to extreme pressure to reduce the Federal deficit, the Bureau of
   Reclamation has insufficient funds available at this time to perform
   basic research on systems such as this which are still in the conceptual
   design stage. However, we wish you continued success in your research
   on desalting systems.

Sincerely,

James R. Graham
Acting Chief, Research and
Laboratory Services Division

238

Addendum (B):

Because it is easier for the belittlers to curse the darkness than to light a candle, it is plausible for the inventor to put himself in their boots and to foresee what they can come up with and then explain it beforehand.

On The Turbidity In The Leaning Tubular Lenses:

The turbidity in the leaning tubular lenses and basins due to algal, bacterial, and other contaminants is an expected corollary. However, using several units insinuates the self-imposing task of putting out of commission one unit periodically for chlorinating and cleansing purposes. This will allow getting rid of the unicellular algae and other microorganisms that are bound to develop and multiply in the unit under such conditions.

Wagdy A. Assawah

On The Difficulty In Making The Leaning Tubular Lenses:

Stumbling over the difficulty of making the suggested leaning tubular lenses shouldn't be a hindering factor preventing us from benefiting from this invention. Using split moulds to make upper halves, then making lower halves and consequently adjoining them will overcome this alleged difficulty. A step yonder, specially suitable for the bee-hive battery, is to make split moulds for the whole radiating juxtaposed leaning tubular lenses. Such rounded conglomerate of tubes will be tailored to fit the different sizes of the hexagonal basins in these composite batteries. Manufacturing the leaning tubular lenses, whether separately or conjointly could be made of durable plastics or could be made of cheaper plastics to be of disposable nature, and thus can be replaced at any time.

On The Economic Feasibility Of The Yield:

Make it first, then improve on the prototype. This was and still is the name of the game for every invention. If we do nothing, we get nothing. And this reminds us of the first car, when a man was ordered to precede the contraption and to keep ringing a bell to alert people and horses from the lurking dangers of the unfamiliar

240

horseless conveyance. Again Orville and Wilbur Wright flew their first powered "aeroplane, *Flyer 1*" in December 1903. Their first flight lasted 12 seconds and the plane went only 120 feet.

Neither the inefficiency of the archetypes nor the expected harms made us abnegate these inventions. Also, the 1956 "ENIAC", the first computer that filled several large rooms, metamorphosed in few years into real powerful laptops. Undoubtedly, these seawater desalting instruments will sooner than later be subjected in their turn to transmogrifatory changes. And the rest will be history, resulting in booming industries. The fact is, no invention ever escaped from being transformed into a better one and from the better to a best.

On Manipulating The Apparatus:

Like any machine, it will need human supervision. And I dare say, it will be of a minimal nature. The sun will provide the free energy needed for evaporation. The windmill will provide the free energy for pumping the seawater to both the inclined tubular lenses and to the basins. The latters are provided with outlets to get rid of excessive seawater, yet they are kept replenished with continuously running seawater to avoid salting out. Last but not least, it is needless

241

to say that such apparatuses are meant to be used mainly in arid zones of a subtropical nature, where the sun shines day in and day out and where moderate wind velocities prevail. And best of all, the desalting batteries are environmentally friendly.

For a noteworthy remark, the shortage of potable water, entices the compassionate to think alike. Accordingly, Mr. Dean Kamen is on the verge of launching an exciting invention for purifying polluted waters used by many in the subcontinent of India and other underdeveloped countries. There from, sparing many lives from dying either by microbes or by toxic waste.

IV. MANUALLY DRIVEN WATER PROPULSION DEVICE*

By

Wagdy A. Assawah	Nader W. Assawah
Patent Number,	5,743,772
Date of Patent,	April 28, 1998

A Concise Recountal:

Man has always endeavored to move from one place to another using simple gadgets that can be worked by his bodily strength. In 1839, Kirkpatrik Macmillan invented the bicycle. In 1867, that is twenty-nine years later, his bicycle went into use when Pierre Michoux manufactured it. In 1992 my family and I visited the Space Museum in Washington, D.C. Exhibited was a pedal powered plane. It meant that from the epochs of Icarus, man yearned to fly in air, depending only on work exerted by his muscles. At this point in

---

* N.B. This is the only patent for which the rights for its protection are still in effect. And, since patented, the submitted manuscript is not shown here. For further readings please consult the published patent.

time, a question crossed my mind. Why a cousin of the earlier two devices, that can help man to move on water, lagged behind till the dawning of the twenty-one century.

Together with my son "Nader", we managed as explained by the accompanying figure entitled "The Muscle-Powered Pedal Trilogy" to come up with what we called "The Hydrobike". It was a simple manual device that assists the user to skim on water with ease. Our patented invention was none other than a human being fitted with a manually driven outboard motor coupled in epitomized simplicity. Though lucid, yet the idea seemed to be the renitent simplicity that confounded the earlier inventors who came up with elaborate contraptions. In potentiality, our adjunct water-propulsion device is an exoskeleton apparatus that can enhance man's prowess in open bodies of water. Undoubtedly, the motorized version has a James Bond appealability.

Addendum A:

With a legalized patent, my son and I knew that we can not sit on the glory of having a patent. We would not sell our invention by standing still. We have to publicize it to show its uses and capabilities. The following is a selected promotion letter that was sent to the concerned in the hope of creating a bonanza to the country and to the entrepreneurs.

The Director:

The Named Enterprise:

Dear Sir,

The concept of the propeller was known for 100 years ago or more. It was and still is used to fly airplanes, lift helicopters, provide air cushions for hover-crafts to buoy on water bodies and skip nimbly on land, to run ships, to move launches, to jostle the flat-bottom crafts of Florida everglades... etc. Yet incorporating a propeller to humans lagged behind till the winding up of the 20th century when my son and I came up with a plausible patented solution. Our water propulsion device is none other than a human being fitted with an outboard motor coupled in epitomized simplicity. Please see the accompanying figure entitled, "The Hydrobike is here".

Wagdy A. Assawah

The propeller in our water propulsion device is either spinned manually hence the name, "The Hydrobike" or by a multispeed 3 forward and one reverse electric motor; hence the name, "The Human Torpedo". Comparable to a fish, the human torpedo has a propeller for a tail and free hands for pectoral fins. Not worrying about his means of propulsion, our merman can put, so to speak, his "pectoral arms" to rescue the drowned, to collect specimens from the sea-bed, to photograph the ecosystems of coral reefs...etc. The device provides a new method to incorporate a propeller in juxtaposition of the heels of the user to propel himself on water surfaces. Simply we are making mermaids and mermen from the users of our innovated device.

Hearing from you will be greatly appreciated.

Respectfully Yours
Wagdy A. Assawah

The Hydrobike is here!

1-The kite enables you to share the air with birds.

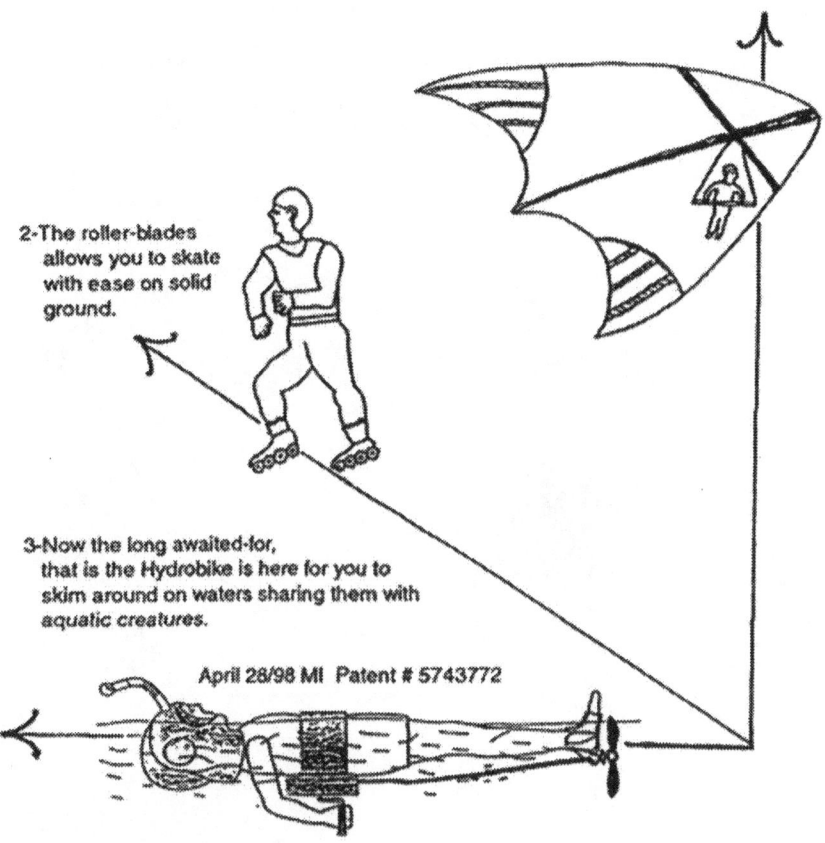

2-The roller-blades allows you to skate with ease on solid ground.

3-Now the long awaited-for, that is the Hydrobike is here for you to skim around on waters sharing them with aquatic creatures.

April 28/98 MI Patent # 5743772

So come "Ye All," young and old to flex your muscles and to experience the excitement of this new water-sport in pools, river estuaries and inner lakes.

Best wishes.

The Muscle-Powered Pedal Trilogy

Wagdy A. Assawah and Nader W. Assawah

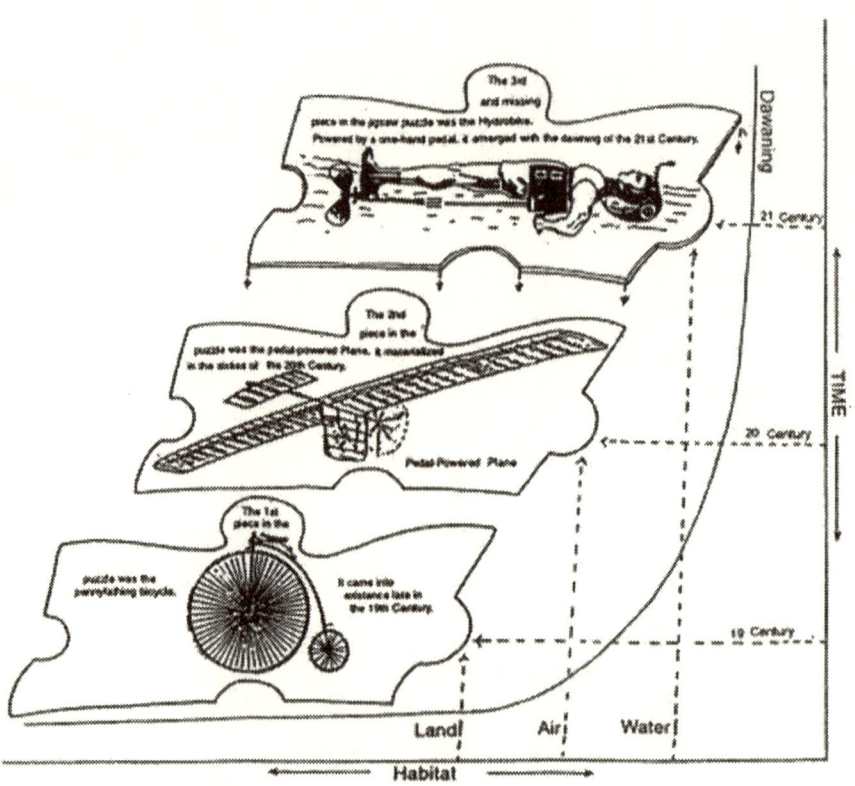

The Curve showing the correlation between a habitat and the time it succumbed to a muscle-powered pedal gadget.

First came the pedal powered pennyfarthing bicycle to run on land in the 19 century.

Second came the pedal powered plane to fly in air in the sixties of the 20th century.

The inventors surmised that if there are pedal gadgets for land and air, there should be a comparable one for water.

Convinced of a third missing piece in the jigsaw puzzle, they invented the Hydrobike. It was systematic deliberate planning that brought this missing link to light to establish the Muscle-Powered Pedal Trilogy.

Wagdy A. Assawah
Addendum B:

Reading about Mr. Michael Moshier and his personal flying machine in *"The Philadelphia Inquirer"* of February 11, 2001, my son and I knew all along that our water propulsion device displays comparable characteristics ascribed for Mr. Moshier's flying vehicle. In his vehicle, the pilot will take off and land like a helicopter. He would stand on two footrests, lean on a sliding backrest, and grip two handles that control the tilt and speed of two hula hooped ducts that blow air for lift off and for transporting its user in a standing position.

The Solo Trek is one of many projects being considered for a program at "The Defense Research Agency" called "Exo-Skeletons for Human Performance Augmentation". The table entitled, "Compare and Contrast" is for showing the similarities and dissimilarities between the forwarded "Water Propulsion Device" and "The Flying Vehicle" of Mr. Moshier.

COMPARE AND CONTRAST

Name of Apparatus	The Flying Vehicle	The motorized Water Propulsion Device
Placement	Exo-Skeletor	Exo-Skeletor
Haulage	Solo Trek	Solo Trek
Purpose	Human Performance Augmentation	Human Performance Augmentation
Medium of Activity	Open Skies	Open Water bodies
Powered by	Fossil Fuel	Electric Motor
Task	Help soldiers to get around obstacles	Help a navy seal to move stealthily
Safety	Disastrous if it malfunctions in air	Not life threatening if it malfunctions at sea
Awareness of Invention	Attracted the attention of the Military	Wrote twice, couldn't lure the Navy's interest.
Accountability	Millennium Jet Inc. of SunnyVale, California	Patent number 5,743,772 of Assawah *et al*
If Combined	We will be having "The Triphibious Man".	

Used, the Navy need not send a dolphin to do a Human-Torpedo's job.

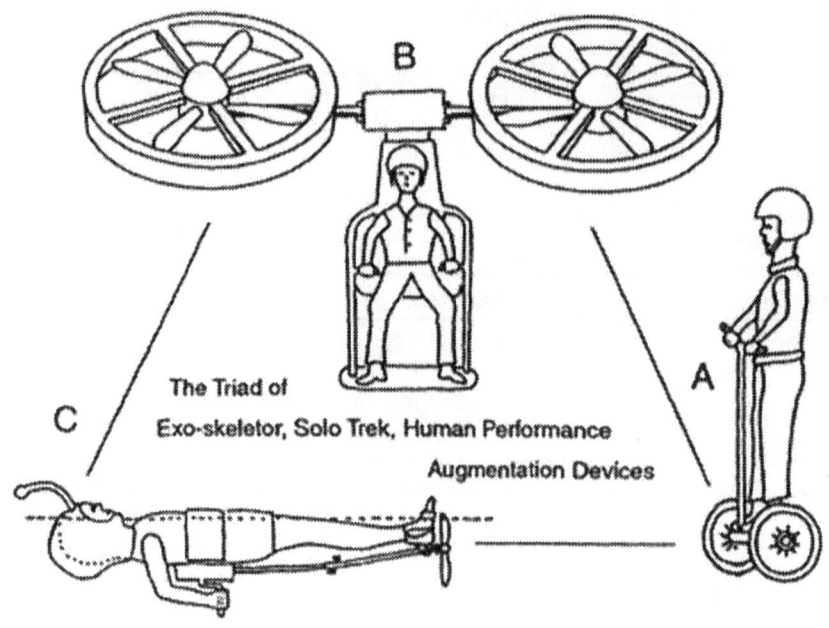

The Triad of
Exo-skeletor, Solo Trek, Human Performance
Augmentation Devices

The sketch is for explaining the shared innate traits of the above three inventions. Surprisingly, one can sense a measure of telepathy for the embraced concepts, but not for the specificity of the medium. Don't you agree that this was a blessing in disguise.

For Mr. Dean Kamen, the medium was land, and he came up with: A

"The Segway Human Transporter"

For Mr. Michael Moshier, the medium was air, and he came up with: B

"The Flying Vehicle"

For Mr. Assawah and son Nader, the medium was water, and they came up with: C

"The Water Propulsion Device "W.P.D."*

The "Human Transporter" was released on December 03, 2001 and was greeted with a bang. The "Flying Vehicle" is awaited for with great expectations.

However, the W. P. D. that was patented on April 28, 1998, was met with a whimper. This does not shame us. It shames the industry who imperceptibly missed that it could be a prized swan. Now leaving the Asian industry to manufacture it, and to prosper from it, while our economy is in dire straits, will be preposterous.

---

* *This W.P.D. deserves to be enjoyed by America, to receive recognition and to hit the list of who ever can make it, or shall I say hit the "Dean's List."*

Wagdy A. Assawah

V. PROTECTED HELIPORTS TO SAFEGUARD COMMUTERS

Filing Date: 12/29/98   Application No., 09/223,369

Foreign Filing License   Granted 01/26/99

A Concise Recountal:

The privacy of Heads of State, dignitaries, and celebrities, is a free game for peeping toms and press hounds spying on their movements, or worst shot at by menial snipers. Snapshots taken by incessant paparazzi are then sold for the highest bidder. Such unwelcomed intrusions resulted in the regrettable death of Princess Diana. And up till now, the world is still missing her humane efforts in many endeavors especially in dismantling land-mines. Thence, for insuring the safety of the esteemed leaders, the rich and the famous, protected heliports with movable hedges that can be operated by remote control in a manner reminiscent to car-garage doors, were deviced.

Hovering in air with the intent to land, the commuter clicks on his remote control to lower the hedge surrounding his rounded landing platform to allow for safe landing. With the copter coming to rest, the commuter while still inside his copter clicks again to raise the hedge. This will allow him to disembark and to leave his copter under the protection offered by the raised hedge. Together with a covered escalator or a covered truss bridge which puts the heliport in continuum with his home, complete privacy and safety are thus consummated.

For departure, the commuter performs the procedure in reverse. He emplanes with the hedge raised. Once inside his copter he clicks his remote control to lower the hedge for a safe take-off.

Wagdy A. Assawah

ON HOW I GOT CONCERNED OVER THE HELIPORT CONCEPT

Coming to America, I decided to make a difference and to leave behind to my new country a legacy that can add to America's prosperity. For brevities, the letters cited here-in, to both the Honorable Mayor Ed Rendell and *the Philadelphia Inquirer,* explain it all. These letters show that I have suggested to the Mayor to build a characteristic land mark to be the pride of America in general and to be the superbity of Philadelphia in particular.

The idea entailed a colossus, for which I gave the name, "The Usalisk", a name rhyming with the famous Egyptian "Obelisk". This coined name took up "U.S.A" as its first syllable and "lisk" as its second syllable.

Our forefathers associated themselves with the sublime. Shouldn't we follow in their footsteps! Their awesome super-structures comprised, but not to exclusion, the following:

- The Gateway Arch of St. Louis
- The Golden Gate suspension bridge
- Mount Rushmore National Memorial

- The Hoover Dam on the Colorado River.

- The Panama Canal, a short cut between the Atlantic and the Pacific Oceans, allowing the biggest ocean liners to pass through.

- The Space Needle of Seattle.

- Envying Seattle, the next as disclosed by *the Philadelphia Inquirer* of December 14, 2003, will be The Tacoma Spire.

Thereupon, the making of the Usalisk, that was suggested by the author nearly ten years to date, will keep a cherished legacy alive. Also bearing with the calling, will make our contemporaneous generations worthy of their ancestry.

This unique Philadelphian land mark can either be reached by elevators servicing the tower or by helicopters landing on high up heliports. Henceforth, the above mentioned brought the idea of protected heliports to my attention. Enticed, I decided to detail this mega-structure in an application to the Patent Office as shown in the following topic, entitled:

"Protected Heliports To Safeguard Commuters"

Wagdy A. Assawah

The Honorable Mayor

Edward G. Rendell

City of Philadelphia.

August 20th, 1994

Dear Sir,

It is high time for the City of Brotherly Love to have its own soaring monument. Surely every one in our great City envy, to name a few: Paris, Seattle, Moscow, Sydney, Rome, Cairo, Las Vegas, for having their own characteristic land marks.

The ultramodern obelisk for Philadelphia will be a continuous reminder that America started here. It will be an enchanting touristic attraction for peoples from different parts of the world to visit our City and to enjoy the panoramic view from the revolving "Sky Capsule". It will be a Mecca for movie directors to shoot pictures, and for entrepreneurs to establish different business and entertainments. Above all, it will be a magnificent time capsule, recording the achievements of the generations of the twenty century and a noteworthy gift to the generations of the dawning twenty one century.

Though, I may not be a certified architect, yet with the basic concepts that I know of, I took upon myself to make a diagrammatic representation of what this land mark should look like. The project may be a financial burden, yet let us hope that leading institutions (i:e the Philadelphia Inquirer) will volunteer to launch donation campaigns to raise the needed funds. Needless to say, the execution of this land mark will create desperately needed jobs. And once completed, its beneficial ripple effect will be a bonanza for big and small business all like.

Wagdy A. Assawah

Dear Mayor Ed, the unchallenged place of Philadelphia in both Historical America and Contemporary America, strongly justifies the construction of this needed monument. I sincerely feel that you, being at the head of this great City, should be the man to take the initiative to inaugurate this land mark and to take the credit for its completion.

Sincerely

Prof. Dr. Wagdy A. Assawah*

---

*Sworn and Subscribed before me. This 24th day of August 1994
Joan Scisco Nicolette, Notary Public Falls Twp. Bucks Country
My Commission Expires April 7, 1998*

August 24th, 1994

The Editor

The Philadelphia Inquirer

Philadelphia City

Dear Sir,

I chose your illustrious institute to forward this letter to The Honorable Mayor Ed Rendell. So please take the time to read it and you will find out that the suggested project is a many splendid thing. To mention a few, allow me enumerate some.

- The venture is going to be a valuable opportunity for badly needed jobs.

- It will be an alluring colossus summoning visitors to pay tribute to our Great City.

Wagdy A. Assawah

- It will be a bonanza for entrepreneurs to establish restaurants, motels, arcades, fun-parks, circuses, fast-food stands, memorabilia boutiques…etc. As such it will be a powerful stimulant to our sagging economy.

- It will be a noteworthy land mark needed for our prestige.
- It will be a time capsule for safe-keeping our achievements.

- It will be an elegant gift to the generations of the dawning twenty one century.

As you see Sir, it is up to you to blow the whistle on this project or to blow it out. But my intuition is that you will not give it the cold shoulder simply because it is a win win venture.

Sincerely

Wagdy A. Assawah*

---

*Sworn and Subscribed before me This 24th day of August 1994
Joan Scisco Nicoletti, Notary Public Falls Twp. Bucks County
My Commission Expires April 7, 1998

Addendum

Like a Cambridge University Collegiate of him, "Prince Charles", who strived to give London a face-lift, the author suggested to the Honorable Mayor Edward G. Rendell in 1994 to build a towering colossus to dominate the skyline of Philadelphia Together with "The Liberty Bell" both memorials will publicize that America started here, in historical Philadelphia.

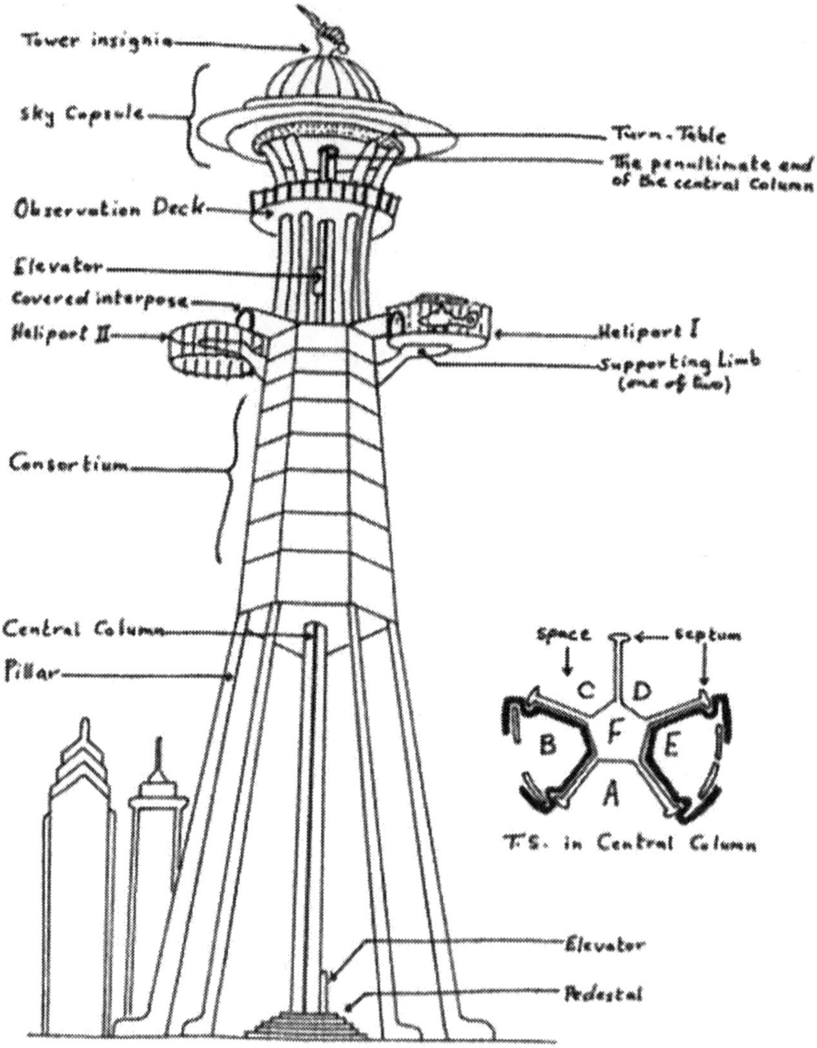
Tower insignia
Sky Capsule
Turn-Table
The penultimate end of the central Column
Observation Deck
Elevator
Covered interpose
Heliport II
Heliport I
Supporting Limb (one of two)
Consortium
Central Column
Pillar
Space
Septum
C
D
F
B
E
A
T.S. in Central Column
Elevator
Pedestal

Philadelphia's Usalisk by Wagdy A. Assawah

The contrived is from a letter to the Honorable Mayor Edward G. Rendell sent to him on August 24, 1994. As such, Philadelphia will be the first City in the world to put its visitors on a lofty Tower by Copters.

Legend Of The Displayed Drawing

The Tower's Insignia:

Showing a hand holding a lightening spark emblematizing our double edged cognizance. So, will it be power that can destroy our world time and time again? Or will it be power that can bring prosperity to all living creatures?

The Sky Capsule:

The dome of this capsule is covered with photoelectric cells to generate the needed power for its rotatory movement.

The Turn Table:

This component is fixed to the uppermost ends of the pillars and carry on its upper side, roller bearings to allow the sky capsule to move in a rotatory manner.

The Central Column:

Its penultimate end furnishes a pivot around which the sky capsule rotates, while its lower most end rests on a pedestal to accommodate the elevators servicing the tower.

The Observation Deck:

A hedged open plateau for lounging and sight seeing.

The Covered Interpose:

A short tunnel-shaped structure to ensure a safe walk from the suspended heliport to the roof of the consortium.

Heliport I:

Showing a landing copter with the protective hedge raised in place to prevent the commuters from falling off the edge of the discoid platform.

Heliport II:

Showing the protective hedge hanging down from the platform's circumference clearing it for a coming copter.

The supporting limbs:

Each heliport is held in place by two supporting limbs coming out from the upper most part of the consortium.

The Consortium:

A Multistory pentagonal structure-could be hexagonal too-built halfway up the tower. Every tier in it will be a museum devoted for displaying our advancements in each and every endeavor.

The Pillars:

These provide the upright support for this superstructure.

The Pedestal:

It is for supporting the lower end of the central column, where from the visitor get on the elevators to reach the different components of the colossus.

The Elevators:

These are five in number for either going up or going down. They are situated in corresponding lacunar spaces created in the central column as shown in the accompanying transverse section T.S. in this pivotal element.

Wagdy A. Assawah

The Transverse Section In the Central Column:

The T.S. is showing core F and five septa as partitions separating the lacunar spaces A, B, C, D and E. These spaces extend from the pedestal to the Sky Capsule. Each space is designed to hug an elevator. In B the embraced elevator has its bivalvate door opened, while in E the embraced elevator has its bivalvate door closed.

PATENT APPLICATION

OF

WAGDY A. ASSAWAH

FOR

PROTECTED HELIPORTS TO SAFEGUARD

COMMUTERS

Background Field Of Invention:

Spying on celebrities, dignitaries, heads of state…etc. with now-a-days sophisticated equipments created the need to safeguard their privacy. Also, to fend off would be assassins, hence protecting their lives too. The suggested protected heliport was thus designed with this aim in mind. Built in their backyards, these heliports will protect the affluent and the influential on their way out or on their way in, from being spied on by peeping toms, incessant paparazzi, or worst menial snipers.

Wagdy A. Assawah

Background-Prior Art:

Apart from protecting the privacy of helicopter commuters, the use of the protected heliports may ease up the traffic jams of today. Earlier solutions, such as "Horizontal Elevators" to link homes and offices, "Pneumatic trains" "Flying Cars," "Rocket Belts" or "Flying Platforms" proved to be impracticable and did not gain recognition.

The presented here-in art, i.e., "The Protected Heliport" have none of the disadvantages of the earlier gizmos. In the inventor's opinion the protected heliports will, so to speak, pave the skies for many to move freely in the coming millennium without being inconvenienced by the traffic jams or the press hounds. Its use will definitely facilitate living on hundreds of thousands of scattered uninhabited islands bordering the shores of Northern Europe, East Asia, the Americas, and in particular the islands of the convoluted shores of the Scandinavian countries, Canada and Alaska.

Objectives And Advantages:

In a manner similar to accessing one's own car garage, the copter commuter, while in his hovering copter, presses his remote control to lower the protective hedge surrounding the landing platform (Fig. I). Thus lowered, the copter is allowed to land on the heliport. Settling on the platform, the commuter, while still in his copter, presses again on his remote control to raise the surrounding hedge. With the hedge raised in place around the platform, the commuter will be able to disembark and to leave his helicopter in complete privacy. A covered escalator or a covered flight of stairs takes him either to his backyard or to his waiting vehicle. The commuter, for completely concealing his activity, may choose to have a covered truss bridge that puts his heliport in continuum with his home or office.

Thus, with the protective hedge raised, the commuter can embark or disembark his copter safely without being spied on or worst, shot at. Meanwhile, with the protective hedge lowered, the copter will, so to speak, have more elbow room to either land or to take-off with ease.

Hence, the objective of this invention is to provide a new and improved super structure that is practicable and reliable for both avoiding the traffic jams and for protecting the privacy and most probably the lives of their users.

The limitations of the "Horizontal Elevators" and the impractabillies associated with the earlier inventions such as the "Flying Cars," the "Rocket Belts," the "Flying Platforms" can be realized by persons familiar in the art. These earlier inventions neither provided genuine solutions for traffic jams nor provide the treasured privacy and the ironclad safety measures vital to their users.

Consequently, it is the objective of the present invention, to provide a sure method for protecting copter commuters from being spied on or worst, shot at by a sniper. Another objective of the present invention is to provide an easier way to avoid the traffic jams and to spare the time wasted on polluted highways.

Still yet another objective of the present invention is to facilitate dwelling on isolated islands bordering our shores, since taking to the rotor will make them nearer to the sites of financial, academic and other intellectual activities.

DESCRIPTION OF DRAWING FIGURES

The invention will be better understood when consideration is given to the following detail descriptions there-of. Such descriptions make reference to the annexed drawings wherein.

- Fig. I: Showing a medial longitudinal section in a protected heliport supported by four pillars and with a landing helicopter on its platform.

- Fig. II: Showing a top view of the discoid platform used for the take-offs or the landings of helicopters.

- Fig. III: Showing a detailed portion of the platform rim, which caters for the different components of the protective hedge.

Wagdy A. Assawah

- Fig. IV: Showing a prospective view of a T-rail. T-rails are equidistantly located on and firmly integrated with the rim of the discoid platform.

- Fig. V: Showing one segment of the protective hedge. Each segment has a rib acting as a backbone for the two oppositely flanking webs ending with bent, thickened edges.

- Fig. VI: Showing the underside of the heliport, disclosing the radiating levers that work the protective hedge.

- Fig VII: Showing two levers, one in top view and the other in side view. Also displaying 3 transverse sections (T.S.) to disclose inner changes along their length.

- Fig VIII: Showing a vertical section in the cap blinding the lower end of a rib. It also discloses both the make-up of the protuberance ending the lower cap and the mode it adjoins the rib with the lever working it. The lever here is shown in T.S.

274

· Fig IX: Showing a medial vertical section in the discoid platform revealing details of the jackshaft and other components, pertaining to the mechanism operating the protective hedge.

· Fig. X: Showing how the hinged end of a lever adjoins with the sliding metal casing.

· Fig XI: Showing a vertical section in the free ends of two hanging down bars to disclose how a pulley is held in place and at the same time allowed to sway *in situ* around its axis.

· Fig. XII: Showing a side view of the free ends of the two hanging down bars to disclose how the middle part of the lever slides to and fro through the swaying-though held in place-pulley.

· Fig XIII: Showing how the cogwheel is fixed in a penultimate position subservient to the turntable terminating the upper end of the jackshaft.

· Fig. XIV: Showing how the skid wheel, though allowed to revolve, is kept in place by a spile screwed on into the free end of the jackshaft.

REFERENCE NUMERALS IN DRAWINGS

1. The System, representing the preferred embodiment.

2. Discoid platform.

3. Landing Helicopter (= copter).

4. Pillars raising platform.

5. Levers.

6. Rib, a backbone of two flanking webs.

7. Electric Motor.

8. Chain.

9. Cogwheel.

10. Jackshaft.

11. Embossment.

12. Skid wheel.

13. Sliding metal casing.

14. Rim of platform.

15. Exit to covered escalator or truss bridge.

16. Webs, two for each rib.

17.Trapezoid (=fish-tail) groove.

18. Bent thickened edges of webs.

19. Connecting strip for concealing the suture between sister webs.

20. T-rail.

21. Concentric markings for copter to centralize its landing.

22. Neck part of T-rail.

23. Surface of neck part to be firmly fixed to the heliport rim.

24. Rectangular free end part of T-rail.

25. Rib groove to accommodate T-rail.

26. Pocket ending the rib groove for housing the rectangular free end of T-rail.

27. Upper cap encasing a night light.

28. Lower cap prolonged into a protuberance.

29. Stout neck of protuberance.

30. Knob, a globular structure ending neck of protuberance.

31. Pair of bars stemming from the underside of the platform.

32. Pulley held by the free ends of the bars.

33. Hinged end of lever.

34. Hole in hinged end of lever.

35. Free end of lever.

36. Snout ending free end of lever.

37. Tubular burrow occupying core of lever.

38. Ostium of burrow.

39. Blind end of burrow.

40. T.S. in solid part of lever.

41 .T.S. in ostium of burrow.

42. T.S. in lever showing burrow opening onto upper side by a narrow adit.

43.Turntable terminating upper end of jackshaft.

44. Lid covering cavity emboxing turntable.

45. Concentric annular protrusion on the underside of lid.

46. Concentric annular trench in turntable matching protrusion of lid.

47. Roller bearings supporting the turntable.

48. Stubby nubs oppositely sticking out from pulley.

49. Holes in the free ends of the bars for clutching the pulley's stubby nubs.

50. Axis of pulley around which the pulley sways *in situ.*

51. Fish-tail pins wedged in holes made in the nubs to keep the pulley in place.

52. Rivets fixing cogwheel in a penultimate position to the turntable.

53. Free end of jackshaft.

54. Screwed on spile in the free end of the jackshaft.

55.Wheel bearing held by spile to ease the movement of skid wheel.

56. Aperture in the pulley accommodating the middle part of a lever.

DESCRIPTION OF THE PREFERRED EMBODIMENT

With reference now to the drawings, and in particular to Fig. I thereof, the preferred embodiment of the new protected heliport embodying the principles and concepts of the present invention and generally designated by the reference System 1, will be described.

Wagdy A. Assawah

Concepts of System 1:

Specifically, it will be noted that the device relates to a new and improved protected heliport. In its broadest context, the device consists of a raised discoid platform. This platform is surrounded by a hedge, that can be pushed up at will or pulled down at will around the platform, with the help of a remote control in a manner reminiscent to accessing one's own car garage door. The working of the hedge is brought about by an electric motor (7). It either pushes up or pulls down the ribs (6) constituting the backbones of the protective hedge. Both motions are brought about by the remote control in the possession of the commuter. A chain (8) conveys the motion from the electric motor (7) to a cogwheel (9) fixed on the stem part of a jackshaft (10).

Levers (5) and ribs (6), depicted in the figure by continuous lines, are in a hanging down deployment to allow for safe landing of the helicopter (3). Levers (5) and ribs (6), depicted by broken lines, are in a pushed up deployment to form a protective hedge surrounding the platform to allow the commuter to leave his helicopter safely and in complete privacy. As a pre-requisite for privacy and protection, the components of the hedge should be, at least 10 feet high, bullet proof, and light in weight. Being light facilitates pushing the hedge up or lowering it down with ease and with minimum energy expenditure.

Complimentary Components For System 1:

System 1 presented here-in, is complimented by various other figures, namely II, III, IV, and V, to disclose other components and to explain other capabilities.

Fig. II is showing a top view of the discoid platform displaying an exit to either a covered escalator, a covered flight of stairs, or a truss bridge. It also displays concentric markings to help a copter to centralize its landing. However, the emphasis is on the structures at the platform rim (14) and how they cater for the components of the hedge. The rim as presented by Fig. III, is studded by T-rails (20) and is vertically furrowed by trapezoid (= fish-tail) grooves (17). The T-rails and the trapezoid grooves are for keeping the protective hedge tightly hugging the discoid platform.

The T-rails shown by Fig. IV, apart from being equidistantly placed around the platform rim, they are also equal in number to the ribs (6). Each T-rail has a neck part (22) having a length equal to the thickness of the rim and a free flattened rectangular end (24) that extends beyond the lower surface of the platform. The surface of the neck (23) integrates with the rim to furnish a strong bond, meanwhile the extended free end (24) ensures more precise guidance of the rib upon ascending or descending. And when the hedge is fully raised, the flattened free end (24) dwells in a matching pocket (26) situated at the lower end of the rib groove (25).

Concerning the protective hedge, it is made of several units. Fig. V discloses the structure of one of these units, showing that it has a mid rib (6) acting as a backbone to the unit. The rib displays a longitudinal T-shaped rib groove (25) meant to accommodate the T-rail (20). Both ends of the rib are blinded by superimposed caps to prevent the rib from dislocating from the T-rails upon reaching its highest deployment, or upon reaching its lowest deployment.

The upper cap (27) blinding the upper end of the rib accommodates a light bulb to aid in night-time landing. The lower cap (28) blinding the lower end of the rib protrudes into a protuberance differentiated into a stout neck (29) and a globular knob (30). The latter is for fitting into a burrow (37) occupying the core of the free end of lever (35), as will be dealt with later on.

On both sides of the rib (6) there are two flanking webs (16). Each web ends with two longitudinally thickened bent edges (18). In T.S., the thickening is triangular in shape. And as the rib is meant to hug the T-rail, the thickened edges (18) are meant to lodge in the trapezoid grooves (17). Thus, when two edges from two sister webs come to meet longitudinally in a trapezoid groove, the thickness of the two will be bigger than the opening of the trapezoid groove. This arrangement will prevent the edges of the sister webs from dislocating from the platform rim, thus ensuring the continued encirclement of the platform with the protective hedge. To further exact the continued juxtaposition of the thickened paired edges and to conceal the suture between them, a longitudinal connecting strip (19) is riveted on the two neighboring edges of the sister webs.

The Underside Of System 1:

The underside view presented by Fig. VI is showing how the different components of System 1 co-ordinate with each other in the preferred embodiment.

Apart from indicating where the pillars (4) meet with the underside of the platform, and where the electric motor (7) is located, the figure reveals that the levers (5) diverge in a radiating fashion towards the platform rim (14). And that the hinging ends (33) of the levers originate from a sliding metal casing (13). This latter structure encircles the stem part of the jackshaft (10), presented here by the inner circle in the given figure. On its way to the rim (14), the middle part of each lever passes through an aperture made in a pulley (32). The free end (35) of the lever terminates under and bonds with the protuberance of the lower cap (28), blinding the lower end of the rib.

Each pulley (32) is held in place by a pair of hanging down bars (31) stemming from the underside of the platform. Not only does the pulley support the middle part of the lever, but also allows it to slide to and fro in its aperture (56). Though this movement is limited, yet it necessitates the swaying of the pulley *in situ* around its axis (50).

It may be needed to indicate that the figure, though showing the electric motor and the chain that conveys the motion to the cogwheel (9), yet the latter is not to be seen. That is because the cogwheel is situated under the wider sliding metal casing (13).

Levers, As Arms Working System 1:

The levers, as shown by Fig. VII, are long, uniform cylindrical rods. Each lever has a hinging laterally compressed end (33), displaying a hole (34) in it for adjoining with the sliding metal casing (13). The lever's free end (35) terminates with a turned up snout (36). This snout is for the lower cap blinding the rib to rest on and to be prevented from going any further.

Each lever accommodates a burrow (37) occupying about 2/3 of the lever's length. The burrow, in one direction ends with a blind end sub terminal to the snout (36). In the other direction, it opens with a relatively wider ostium (38). Running along the upper side of the lever-as oriented by the turned up snout and the flattened hinging end-the burrow opens by a narrow adit (42) starting from the ostium and ending penultimately to the snout.

Fig. VIII elucidates how the protuberance of the lower cap hooks up with the lever (5) presented in the figure by a transverse section. The globular knob (30) of the protuberance is shown

Wagdy A. Assawah

occupying the burrow, while the stout neck (29) of the protuberance is clasped by the narrow adit (42) of the burrow.

This arrangement of a bigger globular knob than the narrow adit of the burrow, holds the rib loosely attached to the free end of the lever (35), yet fastly anchored to it. The globular knob and the stout neck of the protuberance are meant to slither to and fro, the first in the burrow and the latter in the narrow adit. Yet, the distance covered by the two components of the protuberance, in their seemingly to and fro trajectory, will only be confined to the free end of the lever (35) adjacent to the snout.

ASSEMBLING THE DIFFERENT COMPONENTS OF THE HELIPORT:

Built according to the required specifications to handle a particular helicopter, the protected heliport will be primarily composed of a discoid platform (2) raised on four or more pillars (4).

The preferred embodiment (Figs. I and IX) is provided with a medial specially shaped rounded compartment overlying the central embossment (11). As for the rim (14), that will cater for the components of the protective hedge, it is provided with studded T-rails (20) and vertically furrowed by trapezoidal grooves (17). Both the T-rails and the trapezoidal grooves, as per the offered design, are equidistantly spaced and made of durable metal to stand the wear and tear of continual usage.

Hooking Up The Protective Edge: (Figs. I, II, III, IV, AND V)

As perceived by the given, the rib (6) and its two flanking webs (16) constitute a working unit of the hedge. These units are first blinded at their lower ends by the lower caps (28). They are then pushed up the rim (14) one by one, so as the rib groove (25) hugs the corresponding T-rail (20) and the bent edges (18) to lodge into the trapezoid grooves (17). To keep these units in place, their ribs are then blinded at their upper ends by the upper caps (27) encasing night-lights. While hanging down and prevented from falling to the ground by the installation of the upper caps, riveting of the connecting strips (19) can be made to seal the sutures between sister webs.

Installing The Jackshaft And Its Subservient Units: (Figs. IX, X, XIII And XIV)

In a typical embodiment, the first element to go into the central compartment occupying the medial embossment (11) will be the roller bearings (47) meant to shoulder the discoid turntable (43) of the jackshaft. After that, the stem of the jackshaft (10) is lowered into the opening in the middle of the embossment (11). With the turn table resting on the roller bearings, the lid (44) of the emboxing cavity is then screwed on and leveled with the rest of the discoid platform.

On the barren jackshaft stem (10), the first subservient component to be housed on it, will be the cogwheel (9). It is slipped on the jackshaft stem via its free end, and then riveted at a predetermined position penultimate to the turntable (43). Thus fixed in place, the cogwheel is then connected to the electric motor (7) by means of a chain (8). This latter conveys the rotary movements generated by the electric motor.

Unlike the fixed cogwheel (9), the sliding metal casing (13) is intended to journey up and down along the jackshaft stem. This is made possible on account of the screw-threading displayed by the stem part and the oppositely matching screw-threading lining the inside of the sliding metal casing (13). For that, this latter is swivelled on the stem part to be in juxtaposition to the cogwheel. There, it is kept ready for the adjoining hinging ends of the converging levers (5), Fig. VI.

The last part to go on the jackshaft will be the spile (54) sustaining the free revolving skid wheel (12).

Adjoining The Protective Hedge To Levers: (Figs. VII, VIII, IX, X, Xl, and XII)

As shown in the preferred embodiment, each pair of bars stemming from the underside of the platform is designed to clutch a pulley (32) between its two free ends. The pulley, though allowed to sway *in situ,* is kept in place by fish-tail pins (51) wedged in holes made in the oppositely protruding nubs (48).

The protective hedge, that was earlier assembled around the rim (14) of the platform, is now in a free, hanging down disposition.

Wagdy A. Assawah

Levers (5) are then brought to where the hedge is. Thence, in each lever, a globular knob (30) ending the lower cab (28) is inserted into the ostium (38) of the burrow (37). The knob is then allowed to proceed in the burrow for a short distance with the adit of the burrow clasping the stout neck (29) of the protuberance.

The reason behind the shorter distance traveled by the knob, is to leave ample room for threading the hinging end (33) of the lever through the aperture (56) of the pulley (32). Thus threaded, the hinging ends (33) of the converging levers are then pushed further towards the centralized jackshaft to reach and adjoin with the sliding metal casing (13). With the completion of this last piece of work, the protected heliport will be ready to deliver the privacy and the protection expected from it.

OPERATING THE PROTECTED HELIPORT

The drawing in Fig. IX is showing a prescriptive presentation of the preferred embodiment to explain the working mechanism of this superstructure. For the sake of clarity, the role played by one lever only, is emphasized to allow for displaying it in different deployments. Consequently, the collective actions made by all the levers can be conceptualized. Other drawings, shown by Figs. X, XI, XII, XIII, and XIV, were deemed necessary to unravel the gadgetry of the different pivotal parts catering for the end result.

Specifically, it will be noted that manipulating System 1, is totally dependent on the centrally located jackshaft (10). Basically, it is composed of a stem and a turntable (43) terminating its upper end. The stem displays a screw-thread and the turntable displays an annular trench (46) on its upper side. When placed in the cavity made in the central embossment (11), the turntable will be sandwiched between a roller bearing (47), shouldering the underside of the turntable and a lid (44) overlaying the upper side of the turntable. The annular protrusion in the lid (45) matches the annular trench (46) of the turntable. This arrangement of an annular protrusion fitting in a matching annular trench together with the shouldering roller bearing will insure smooth spinning movements and no wobbling in the stem part of the jackshaft. Around that stem, there is the sliding metal casing (13) working the levers emanating from it in a radiating arrangement. The spinning of the jackshaft is brought about by the electric motor (7), which transmits its movement to the cogwheel (9) of the jackshaft via the chain (8).

In conformity with the spinning action of the jackshaft (10), whether clockwise or anticlockwise, as brought about by clicking on the remote control operated by the copter commuter, the internally and oppositely screw threaded sliding metal casing (13) moves either upward or downward. When it goes up, as given in position 13a, the free end of the lever goes down and the rib goes down with it. And when it goes down, as given in position 13c, the free end of the lever goes up and the rib goes up with it. But when it is halfway, as given in position 13b, the lever takes a horizontal posture. In this latter case, the base of the rib does not rest on the snout (36) of the free end of the lever, because the snout (36) will be protruding beyond the base of the rib.

The apparent to and fro movement by the rib on the free end of the lever (35) is made easy by furnishing the lower blinding cap (28) with a protuberance composed of a stout neck (29) and a globular knob (30). The apparent movement causes the globular knob to slither in the burrow (37) occupying the core of free end (35) of lever. Meanwhile the stout neck (29) to slither in the narrow adit (42) of burrow (37). This arrangement of a bigger globular knob than the narrow adit of the burrow, holds the rib loosely to the lever, yet anchored to it.

The spinning of the jackshaft (10) around its axis is designed to shut down when the sliding metal casing (13) reaches its upper most deployment or when it reaches its lower most deployment. This is comparable to what takes place in a car garage door operated by remote control. Yet the lower end of the jackshaft (10) is provided with a loosely revolving skid wheel (12) to allow the sliding metal casing (13) to rest on it when coming to its lowest deployment, 13c.

In going up or down, the ribs together with their flanking webs (16) are guided by the T-rails (20) fixed at the rim (14) of the discoid platform. Thus positioned, the T-rails together with the trapezoid grooves (17) holding the bent thickened edges of the webs, orchestrate the vertical ascending or the vertical descending of the whole protective hedge.

To sum up, the movement of the sliding metal casing (13) as brought about by winding the jackshaft (10), forces the radiating levers (5) to push up or to pull down the protective hedge in a manner comparable to folding or to unfolding an umbrella.

Thus, as readily perceived, the protected heliports besides helping to avoid traffic jams they provide reliable safety and treasured privacy for their users. And while the above description contains many specificities, these should not be construed as limitations on scope of the invention. Many other variations are possible both in size and shape. Also, these heliports can come in twos or threes with one common outlet, as may be needed by sizable airports. Or there can be several heliports arranged around and individually connected to a huge rotunda where the authoritative rulers from different countries, Secretaries of States, financial moguls…etc., can convene in the halls of the rotunda to discuss important issues and to decide on suitable solutions in the absence of eavesdroppers, the paparazzi, or worst hired assassins.

THE PAGES ENCOMPASSING THE DRAWINGS

WITH THE FIRST PRESENTING

SYSTEM 1

AS THE PREFERRED EMBODIMENT OF THE
"PROTECTED HELIPORT"

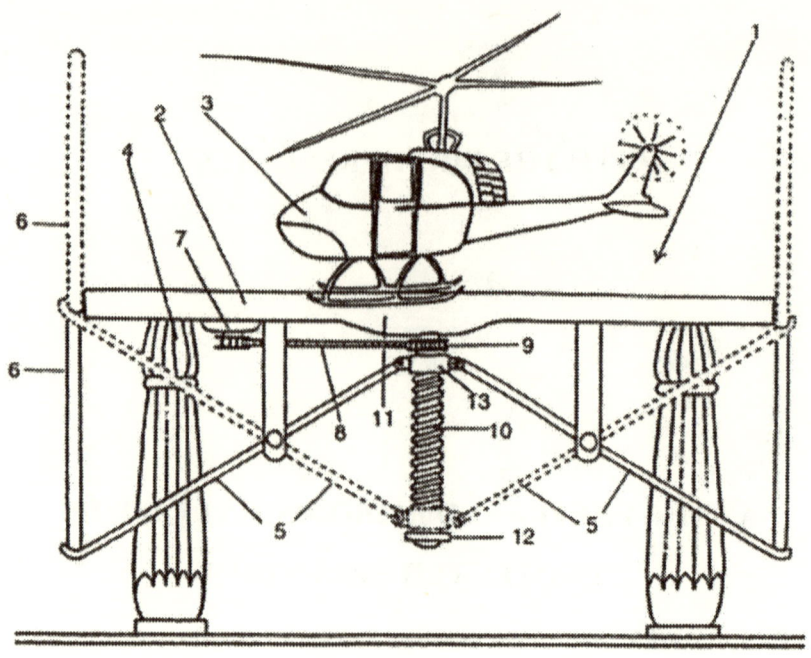

The Heliport

FIGURE I

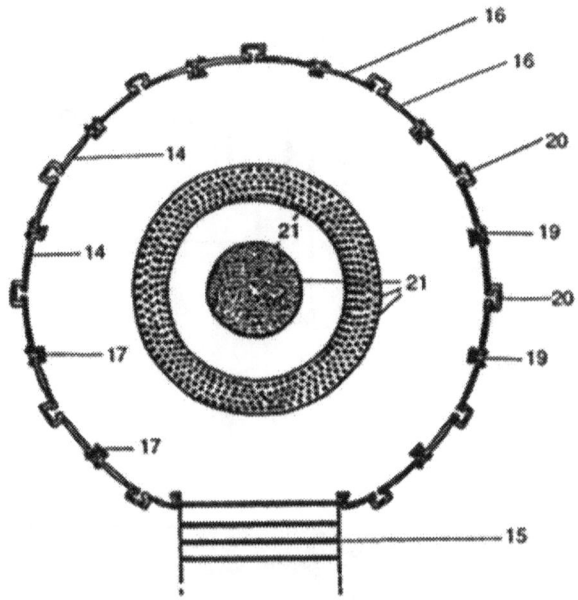

FIGURE II

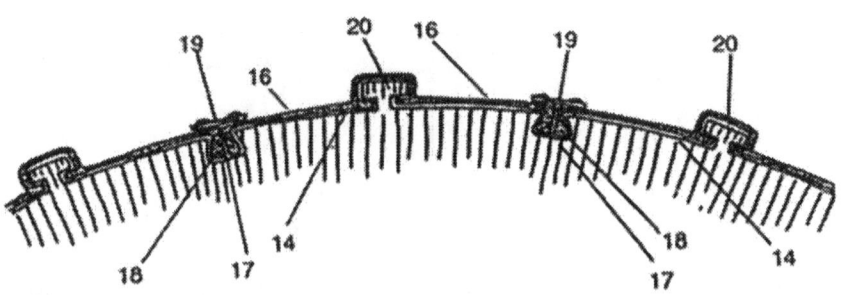

FIGURE III

The Heliport

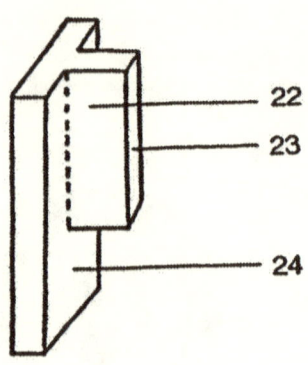

FIGURE IV

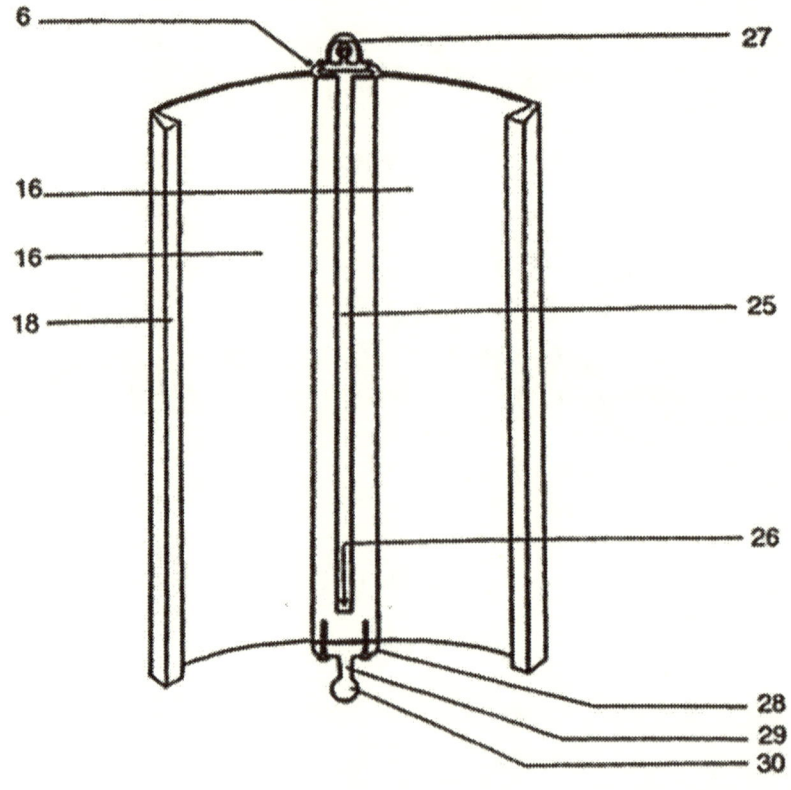

FIGURE V

The Heliport

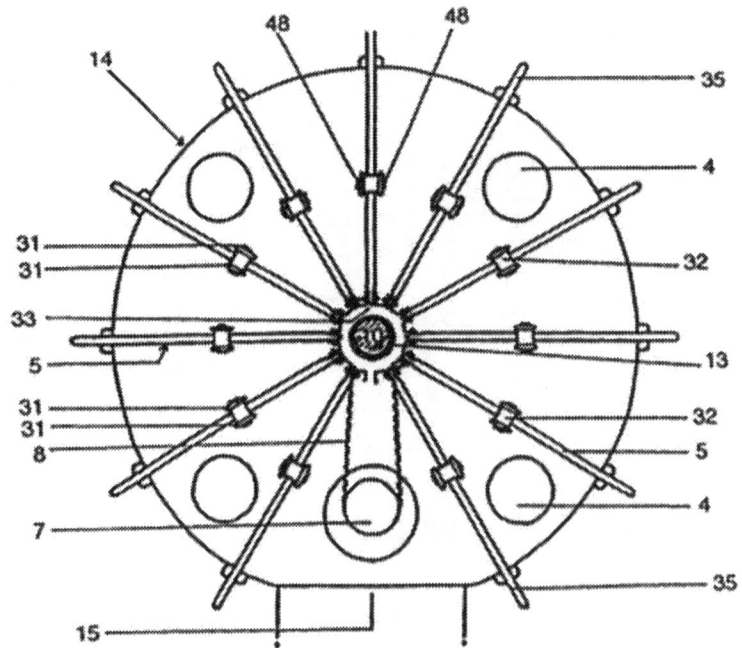

FIGURE VI

The Heliport

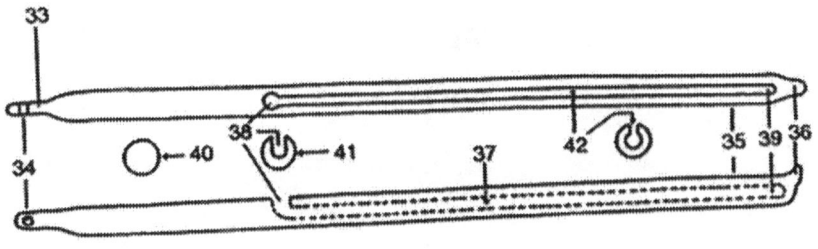

FIGURE VII

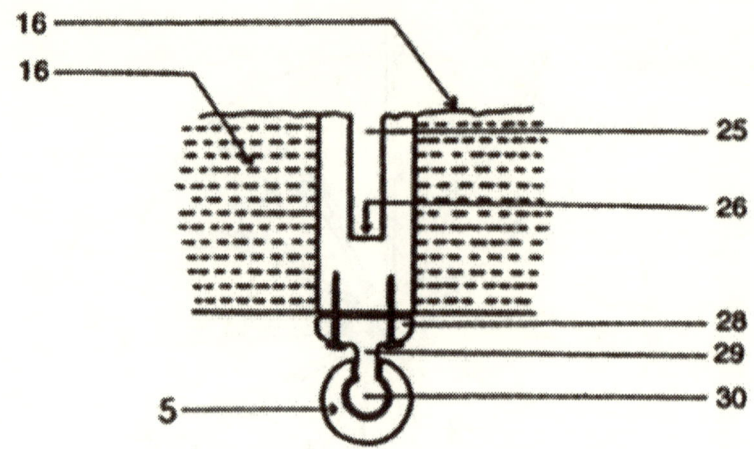

FIGURE VIII

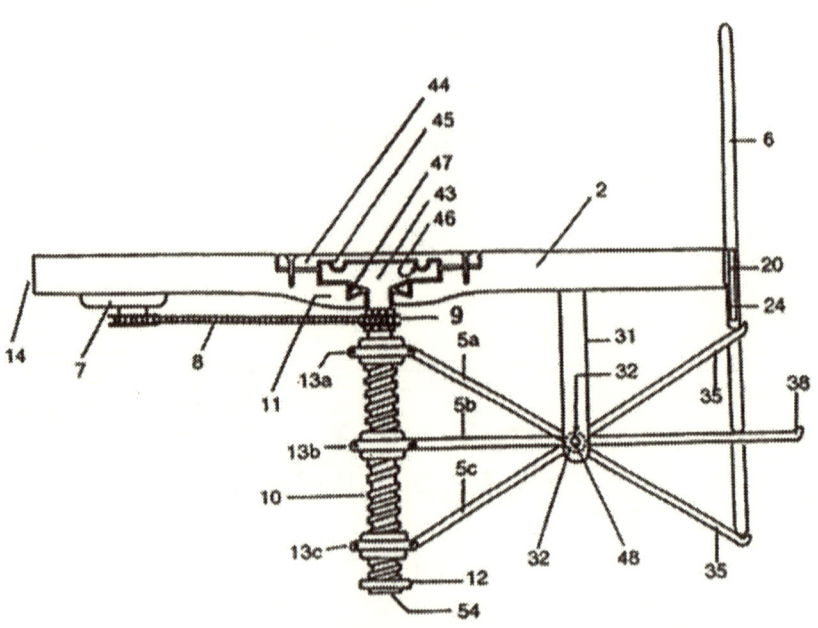

FIGURE IX
The Heliport

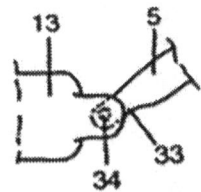

FIGURE X

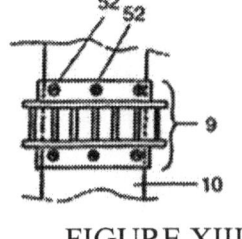

FIGURE XIII

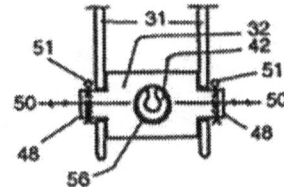

FIGURE XI

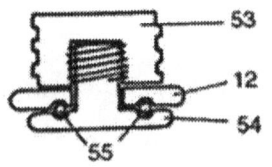

FIGURE XIV

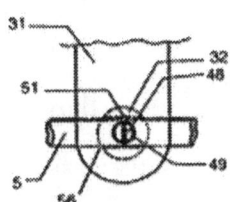

FIGURE XII

The Heliport

VI. (A) EVULSION CLAMPS TO SEVER SKIN FLAPS

Filing date: 3/03/99   Application No.: 9/261,848

Foreign Filing License   Granted: 9/19/99

A Concise Recountal:

"Evulsion Clamps" made of strong yet light materials are designed to sever non-surgically the unwanted skin flaps or creases from the human body. Such flaps could be small like the so called "double chin" or they may be big as the "love handles" or bigger still as the nicknamed "spare tires" skirting the bellies of the very heavy. The E.C. has a pair of bivaluately biting jaws worked by a pair of hafts. The coming together of the biting jaws to sever a flap or a crease will be like the coming together of the sharp edges of a nail-clipper truncating an outgrowing nail. For the patient to endure the pain caused by the pressure of the biting jaws, local anesthesia can be applied topically to ease away the soreness. When the crushing of the biting jaws reaches its apogee, the blood capillaries nourishing the bitten flap will be forced to shut down. Deprived of nutrients and gaseous exchange, causes the trapped flap to starve and to snuff out. Meanwhile the veins on the patient's side will rapidly change course to take over the function of collecting blood and channeling it to the deeper veins of the treated patient. Eventually, the meeting skins crushed under the biting edges of the E.C., will be encouraged to fuse and to form an abscission layer at which the moribund flap separates from the body. It may separate on its own accord; if not it can be helped with minimal medical intervention to do so.

Compared to the yo-yo syndrome of dieting, the perils of invasive surgeries, and the uncertainties of liposuction, E.C.'s are safer deliverers. Because of eventuating minimal medical supervision, together with affordable medical expenses, the E.C. will be a haven for a wider populace to get rid of their scatheful flaps. This non-surgical apparatus introduces what one may call, "The Forced Partial Human Molting". This plucking of flaps, or creases, offers together with a healthier posture for the patient, several other diversified benefits. It will offer an easy way out of the portliness, the root of many maladies. It will lessen the burden on the supporting skeletal system alleviating pains of wear and tear on the vertebrae and other joints. It will lessen the volume of the body mass, hence reducing the workload on the weary heart improving its performance.

Wagdy A. Assawah

Supplementory to the above, on March 3, 2000, a resubmission of my earlier, "Evulsion clamps to sever skin flaps" was forwarded by a patent attorney on my behalf to the P.T.O. It was under the name, "Broad faced instrument and its associated method of use." On August 27, 2002, i:e., after more than 28 months, this previously mentioned resubmission was granted a patent number 6,440,145. Yet this information came at a time when the book was about to be forwarded to the publisher. Hence, instead of including the approved dissertation from the P.T.O., I was inclined to leave the already incorporated manuscript in place. This was specially so, because in it there is more descriptions of certain key components, e.g. the girders lining the inside of the hafts, the floating nuts…etc. Also more explanations on the assemblage and how the different parts fit snugly together to work harmoniously. And most importantly, how to build this particular gadget by the would be manufacturers for achieving the purpose contemplated.

PATENT APPLICATION

OF

WAGDY A. ASSAWAH

FOR

EVULSION CLAMPS TO SEVER SKIN FLAPS

Background-Field Of Invention:

The evulsion clamp "E.C." presented here-in is designated to sever non-surgically the unwanted fatty skin flaps. The suggested E.C. has a pair of bivalvately biting jaws worked by a pair of hafts. The coupling together or the uncoupling of the pair of biting jaws is comparable to working a pair of scissors. While the scissors are worked by the human hand, the E.C. is worked by means of a device spanning between a pair of hafts. And while the scissors cut objects longitudinally, the edges of the pair of biting jaws are meant to meet transversally into a deadlock to snuff out a fatty skin flap and eventually severing it from the patient's abdomen.

Wagdy A. Assawah

Background-Prior Art:

Aware of their appearance, people from all walks of life aspire for younger and healthier looks. Usually efforts to lose weight do not succeed. People who cannot reduce their weight by dieting may then resort to drastic measures. Hence, physicians sometimes try digestive tract surgeries to by-pass parts of the small intestine to reduce food absorption. Or they may try to lessen the size of the stomach by stapling to limit food intake. Such procedures come with high risks of death.

Still many others, seeking to win the -battle of the bulge- find cherished quick fixes in cosmetic procedures. These either involve sculpturing by carving away unwanted skin together with its underlying adipose tissue. Or involve tumescent liposuctioning with cannulae inserted into the body to hoover fat out. Improved liposuctioning techniques involve using high-frequency sound waves to liquefy fat cells before removing them with a lower pressure suction tool. Yet both sculpturing and liposuctioning have their perils and exorbitant costs.

A recent non-surgical procedure is now used in the treatment for cellulite. It involves the use of the "Endermologie Machine". This machine has a suction cup, which pulls a fold of the skin up to knead it between rollers. This was found to homogenize the layer of fat, thus masking the symptoms giving skin deep smoother appearance.

Objectives And Advantages:

As we age, many develop dangling skin flaps. This is noticed upon dieting or due to loss of skin elasticity, as in the elderly. These unwanted skin flaps come with creases lining their underside. If left uncleaned, the creases furnish microcosms harboring assortments of microbial organisms. Some may be saprophytic, thriving on sloughed off epidermal cells and sweat, causing foul smells. Others may be parasitic, causing hazardous health conditions ranging from easy to heal lesions to vicious flesh-eating bacteria. Unfortunately, for economical reasons, the removal of sagging skin flaps is passed on as cosmetic frivolity. Yet their evulsion is justified on account of the ills they cause down the road.

Stored fat though it has no "bad rap," yet it is as harmful as other human byproducts of digestion, namely excrements and urine. These waste byproducts produced upon harvesting energy are poisonous offals. Kept in the human body for longer periods without discharging, they may end the life of the afflicted. Stored fat may be considered as a benign byproduct, yet it has several dark sides, one of which is a shorter lifespan, as evidenced by the higher mortality rates in heavier individuals. Keeping bulky fatty flaps obligates the body to provide them with both the life sustaining nutrients oxygen, and obligates the body to remove the waste products of metabolism. As such, fatty flaps formulate additional labor to both the heart and the supporting skeletal system. Hence an easy and affordable means for disposing of these fatty flaps, as provided by the E.C., can help in reducing the work load on the heart and help in lessening the burden on the vertebrae and other supporting joints.

Up till now, the removal of sagging skin flaps has been carried out surgically under anesthesia. Due to innate fears of surgeries and the expenses bordering on the exorbitant, a simpler, gentler, and an economical method was thought of. Presented here-in, "The Evulsion Clamp" will bite on the skin flap, isolating it from the body of the patient. Where the crushing is exerted, the blood capillaries will be forced to shut down depriving the flap from nutrients and gaseous exchange, causing it to snuff out. The E.C. will also force the capillaries and the smaller blood vessels on the other side of the flap, i.e., the patient's abdomen, to anastomose forming a continuum with the rest of the circulatory system of the patient. The crushing brought about by the E.C., will eventually encourage the formation of an abscission layer at which the snuffed out flap will separate on its own accord, or otherwise with minimal medical intervention. Certainly, there will be pain accompanying the tightening of the E.C., yet local anesthesia can be used to ease the soreness.

The use of the Evulsion Clamp will overcome many of the disadvantages of the contemporary invasive surgical practices.

Compared to by-passing parts of the small intestine, or reducing the size of the stomach by stapling, the severance of unwanted fatty skin flaps by the E.C. is much safer than the previous surgeries, which come with unacceptable high risk of death.

Compared to sculpturing surgeries, these may carve less or do away with more, forcing the patient to go under the knife for more correctional treatments. Meanwhile, the E.C. will obviate the dreaded blood letting and the probable correctional treatments.

Compared to the conventional liposuction, this may leave the skin lumpy due to unequal hoovering of the stored fat. Also hoovering of fat, especially in older persons who lost skin elasticity, may cause the skin to pocket into unwelcomed folds. Meanwhile, with the E.C., severing a fold will leave the treated part, so to speak, clean-shaven.

In fact, all the surgical procedures mentioned above expose the patient to perilous practices with possible risks of death. Though a remote possibility while operating, these procedures may establish ports of entry to deadly blood contagions. They also neither spare the patient from the uncertainties of the results nor spare him from the steep medical bills. Over and above as a convalescent, the patient will be obligated to stay as an intern at a medical facility for post-trauma follow-ups, which again cost a bundle. Meanwhile, with the E.C., the patient will be exposed to minimal risks, if any. He will be able to pursue the treatment in the comfort of his home with minimal medical supervision and consequently, reasonable bills.

Kneading the pocked skin with the "Endermologie Machine" is a non-surgical procedure. And because the "Evulsion Clamp" is also another non-surgical procedure, this earns the E.M. its own right for a separate emulation. The E.M. gives the treated patient a skin-deep smoothness that does not last for long. Having no permanent effect on the cellulite, the patient has to do it over and over again. The E.M. offers no cure, it only masks the symptoms and its continued applications destroy the tenacity of the skin worsening the condition. But with the E.C., it will be cure from the first bite.

The idea of using the E.C. to dispose of unwanted fatty skin flaps will force itself on the stage of therapeutic treatments not only for its advantages over the previously mentioned invasive surgeries, but also emanates from other comparable supporting surgical procedures.

The parallelism between the austere castration exercised in earlier times to sterilize the prisoners of war and slaves is self-evident. Unlike castration, which discontinue the functions of vital sex organs, whether testicles or ovaries, the E.C. will sever dispensable fatty skin flaps, which the human body can conveniently do without.

Circumcision in males of our species is also another practice in which the foreskin is severed under anesthesia to expose the glans. It is mainly done because its underlying crease, if not periodically cleansed, is bound to gather a host of disease inducing microorganisms.

Another canny parallelism lends itself from the severed body parts in accidents. Humans may lose involuntarily a finger, a toe, a hand, or even a limb. Also, in cases of sexual abuse, the victimized may resort, out of revenge, to sever the penis of the abuser. Fortunately, when these ousted parts are secured and kept viable, they can be reattached to the patient. However, apart from the obvious analogy in the involuntary excision of a human part, if compared with the voluntary severance of an expendable fatty skin flap with the E.C., the latter will be a blissful act embraced by the patient.

Consequently, the main object of "The Evulsion Clamp" is to provide a new non-surgical, yet safer technique to sever expendable fatty skin flaps. The offered instrument was also designed with the intention of avoiding the risks of digestive tract surgeries, the uncertainties of sculpturing, and the hazards of liposuction.

Another objective is to have complete control over the fatty flap, which the patient wants to emaciate in the nick of time. Losing weight by dieting is time consuming, with doubtful results if any, and may hurt parts which the patient wants to keep plump and lissome.

Still another objective is to make the E.C. available to a wide populous so they can sever with considerable ease the unwanted skin flaps in the comfort of their homes. And since the procedure requires minimal medical supervision, it will also favor affordable medical bills.

Still yet another objective is to alleviate a host of pains caused by excessive heaviness for untreated portly individuals. Additional weight on the abdomen due to stored fat particularly around the hips shifts the point of equilibrium in the human posture. Adjusting to a new center of gravity induces stresses on the supporting skeletal system. The pressure on the lumber vertebrae causes agonizing back aches. It may also induce prolapsed discs, which upon squeezing the sciatic nerve triggers burning pains shooting down along the thigh. Reducing weight by ousting skin flaps, which can be done more than once, will lessen the burden on the eroding joints, consequently reducing pains of osteoarthritis. And because ousting of fatty flaps reduce the patient's volume, this will lessen the workload on the heart, resulting in a tempered ticker, a healthier survival, and maybe, longer life expectancy.

315

The bonanza of a safer practice to reduce weight, with no blood letting, easier to carry out in the comfort of one's own home, with affordable medical expenses, offering less wear and tear to the supporting skeletal system, reducing the work load on the heart, will encourage the once-bitten to get bitten over and over again.

DESCRIPTION OF DRAWING FIGURES

The invention will be better understood when consideration is given to the following detail description there-of. Such description makes reference to the annexed drawings here-in.

FIGURE 1

Showing a side view of the "Evulsion Clamp." As per the given prospective the E.C. is composed of an upper moiety and a lower moiety. The two moieties are medially hinged together by alternating interlocking segments.

FIGURE 2

A ventral view showing the interior of the upper moiety of the E.C.

FIGURE 3

A dorsal view showing the exterior of the lower moiety of the E.C.

FIGURE 4

Showing: a strut-pin in (A), an L-shaped handlebar in (B), and a swivel-nut in (C).

FIGURE 5

Showing: the E.C. involved in severing fatty skin flaps. Double chin in (A), Grandma wobbly flap in (B) and, love handle in (C). In (A) and (C), the E.C. is presented in side view and in (B) the E.C. is in surface view.

REFERENCE NUMERALS IN DRAWINGS

SYSTEM 1

Denoting the preferred embodiment of the "Evulsion Clamp".

10.   Upper biting jaw of upper moiety.

11.   Lower biting jaw of lower moiety.

12.   Upper haft of upper moiety.

13.   Lower haft of lower moiety.

14.   Alternating interlocking segments from both moieties.

15.   Lip part of upper biting jaw.

16.   Mandibular part of upper biting jaw.

17.   Lip part of lower biting jaw.

18. Mandibular part of lower biting jaw.

19. Edge of upper biting jaw.

20. Edge of lower biting jaw.

21. One of the oppositely flanking appendages extending beyond the edge of the upper biting jaw.

22. Continuous rod in T.S. sojourning in a burrow made in the alternating interlocking segments.

23. Penultimate holes, one in each end of the continuous rod.

24. Fishtail pin wedged in the penultimate hole to keep the continuous rod in place.

25. Super imposed upper girder lining the inside of upper haft.

26. Super imposed lower girder lining the inside of lower haft.

27. Slanting edge of upper girder.

28. Slanting edge of lower girder.

29. Overlapping ridge emanating from the upper haft holding down the slanting edge of upper girder.

30. Overlapping ridge emanating from the lower haft holding down the slanting edge of lower girder.

31. Strut-pin.

32. Upper swivel-nut capping one end of the strut-pin.

33. Lower swivel-nut capping the other end of the strut-pin.

34. U-shaped bays nicked in the double-layered hafts.

35. Screws to pin down the girders lining the inner sides of hafts.

36. Oppositely flanking dowels coming out of the swivel-nut.

37. Nooks oppositely tunneled in the straight sides of the U-shaped bays.

38. Windows opened in the lip of the biting jaws.

39. Windows opened in the mandibular part of the biting jaws.

40.   Nicked notches one on each side of the lower biting edge.

41.   Tubular barrows made in the alternating interlocking segments.

42.   Medial cylindrical bobbin forming an integral part of the strut-pin.

43.   Antipodal limbs of the strut-pin displaying oppositely screw-threads.

44.   Handlebar, a smooth cylindrical rod with a longer arm and a shorter perpendicular arm.

45.   A hole in the cylindrical bobbin to accommodate the handlebar.

46.   The barrel-shaped body of the swivel-nut.

47.   The inside of the swivel-nut displaying screw-threads to tie up with the threading of the antipodal limbs of the strut-pin.

DESCRIPTION OF THE PREFERRED EMBODIMENT

With reference now to the drawings and in particular to Figure 1, there-of the preferred embodiment of the Evulsion Clamp embodying the principles and concepts of the present invention and generally designated by the reference System 1 will be described.

Concept Of System 1:

Specifically, it will be noted that the device relates to a new idea for ousting unwanted fatty skin flaps from the human body by a non-surgical procedure. In its broadest concept, the device consists of a pair of biting jaws, an upper jaw (10) and a lower jaw (11). They are worked by a pair of hafts, an upper haft (12) and a lower haft (13). The biting jaws and their corresponding hafts are hinged medially into a unity by alternating interlocking segments (14). And since the patient is going to carry the E.C. around for about ten days, it should be made of strong, yet light space age materials.

The pair of hafts (12) and (13), like the handgrip of a pair of scissors, can either open wide the pair of biting jaws (10) and

(11) or can bring them to a deadlock. To sever an unwanted flap, the pair of biting jaws are allowed to separate wide apart to engulf the skin flap. With the skin flap entrapped, the biting jaws are brought to a deadlock to snuff out and eventually sever the skin flap from the patient's abdomen. The coming together of the edges, i.e., the upper biting edge (19) and the lower biting edge (20), will be like the coming together of the sharp edges of a nail-clipper truncating an outgrowing nail.

Complementary Components of System 1:

System 1, presented herein, is complemented by other figures, namely Figure 2, Figure 3, and Figure 4 to disclose other details pertaining to its structure. Also to explain certain safety measures when using the device. Figure 1 shows that the "Evulsion Clamp" is made of two moieties held together by more or less medially interlocking hinging segments (14). These two moieties are kept together by a continuous rod (22) sojourning in burrows (41) made in the alternating interlocking segments emanating from the two coupled moieties. To keep the continuous rod in place, it is provided by two antipodal penultimate holes (23) in which fishtail pins (24) are wedged.

On one side of the interlocking hinging segments (14), the E.C. has a bivalvately biting jaw, an upper jaw (10) and a lower jaw (11). Each jaw is differentiated into a lip part (15) and (17) respectively and a mandibular part (16) and (18) respectively. Only the upper lip (15) that extends beyond its biting edge (19) into two flanking appendages (21). These appendages imbricate into two notches (40) antithetically nicked in the biting edge (20) of the lower lip (17). To monitor the squeezed skin flap, rectangular windows (39) are made in the pair of mandibular parts (16) and (18). In Figures 2 and 3, the rectangular windows (38) of the lip parts (15) and (17) are shown narrower than the windows (39) of mandibular parts (16) and (18). This is due to the visual aberration caused by the upward curving of the lip part (15) as shown in Figure 2 and the downward curving of the lip part (16) as shown in Figure 3.

On the other side of the interlocking hinging segments (14), each moiety extends into a haft, an upper haft (12) and a lower haft (13). The distal end of each haft is double layered. This doubling is due to superimposed girders lining the inner surfaces of the distal ends of the hafts. The upper haft (12) is lined by girder (25) and the lower haft (13) is lined by girder (26). These girders have slanting edges (27) and (28) respectively, which lodge underneath overlapping ridges (29) and (30) coming out of the corresponding hafts. This arrangement of a slanting edge wedged underneath an overlapping ridge establishes juxtaposition fitments of girders with the corresponding hafts. More cohesion for the distal double-layered parts of the hafts is affected by screws (35) pinning down the girders to the matching hafts.

Connecting the free ends of the hafts involves the use of strut-pins shown in Figure 4A and swivel-nuts shown in Figure 4C. A strut-pin (31) is a rod like structure with a cylindrical bobbin (42) in its middle part and two antipodal limbs (43). The antipodal limbs display screw threads spiraling in opposite directions, while the cylindrical bobbin has a hole (45) running through its middle part. To work the strut-pin, a handlebar (44) is needed. The handlebar shown in Figure 4B is a smooth cylindrical rod with a long arm and a shorter perpendicular arm. Using the handlebar to turn the strut-pin in one direction causes the swivel-nuts (32) and (33) capping the antipodal limbs to glide nearer to the central bobbin. This, in its turn, forces the hafts to come closer to each other and at the same time obligating the biting jaws (10) and (11) to open wide apart. Using the handlebar to turn the strut-pin in the opposite direction causes the swivel-nuts (32) and (33) to distant themselves from the central bobbin (42). This in its turn forces the hafts away from each other and at the same time bringing the edges (19) and (20) of the biting jaws to a deadlock. These two contradictory movements of the swivel-nuts are made possible on account of the oppositely spiraling screw threads displayed by the antipodal limbs of the strut-pins. As shown by Figures 2 and 3, the presented E.C. has three strut-pins to either open it or to close it. Certainly, bigger skin flaps will need longer E.C.s with more strut-pins to work them.

The swivel-nut, Figure 4C, has a barrel-shaped body (46). Its inside exhibits screw threads (47) that tie up with the screw threads of the antipodal limbs of the strut-pin. Each swivel nut has two oppositely flanking stubby dowels (36). These are made to fit snuggly into nooks (37) tunneled into the straight sides of cutaway U-shaped bays (34) made in the double layered part of the hafts. The swivel-nuts while sliding on the antipodal limbs of the strut-pin are supposed to follow a slightly curved trajectory. Yet they are obligated to slide vertically in a straight path. This was made possible by the help of the oppositely flanking dowels, which allow the swivel-nuts to adjust by swaying *in situ* around the common axis of the dowels.

Assembling The Evulsion Clamp:

The first parts to be assembled will be the two moieties of the E.C. At that point in time, the hafts (12) and (13) should be without their lining girders (25) and (26). The two moieties are then bivalvately coupled by engaging together their alternating interlocking segments (14). Thence the continuous rod (22) is inserted into the burrows (41) made in the interlocking segments to establish a working hinge. With the continuous rod in place, the fishtail pins (24) are then wedged in the penultimate holes (23) to hold it in place.

Strut-pins (31) equaling in number to the U-shaped bays (34) are then capped with swivel-nuts (32) and (33), one on each antipodal limb. The strut-pins together with the screwed on swivel nuts are then placed to span between the corresponding U-shaped bays (34) made in the upper and lower hafts. A tight contingency of the flanking dowels (36) with the half-tunneled nooks (37) made in the uncladed hafts will most probably need fine adjustments. This is done by spinning the strut-pins around their axes till a thorough contact is affected between the flanking dowels and the half-tunneled nooks made in the uncladed hafts. This will be manifested when a complete closure of the biting edges is evident.

The last parts to go on for completing the assembly of the E.C. will be the lining girders (25) and (26). Their slanting edges (27) and (28) are shoved under the overlapping ridges (29) and (30) of the hafts. Also their half-tunneled nooks are paired with the half-tunneled nooks in the corresponding hafts, thus enveloping completely the flanking dowels coming out of the swivel-nuts. To secure a firm juxtaposition of the now bilayered hafts, the lining girders are then pinned down with several screws (35). With the completion of this last step, the E.C. will be ready for use.

Operating The Evulsion Clamp:

To sever an unwanted fatty skin flap, one has to bring the pair of biting jaws to a deadlock. The strut-pins of E.C. have to be worked in such a way so as to ensure continuous gradual and equally distributed pressure along the two lengths of the biting edges. To do this, each strut-pin is turned once starting with a first and ending with a last. As such this will constitute one round of twisting the strut-pins. To bring about a complete closure of the biting jaws, consecutive rounds of turning the strut-pins are made repeatedly till a deadlock is reached. As shown by Figure 2 and

Figure 3, the presented E.C. has three strut-pins to either open or to close it. However, bigger skin flaps will need longer E.C. with more strut-pin to work them. The squeezing of the skin can be executed according to the patient's tolerance. It could be made either in one session or over comparatively longer periods of time. Undoubtedly squeezing a skin flap will be painful to the patient yet the soreness can be eased by localized anesthesia.

Apart from the stepwise squeezing, the trapped skin flap should not be allowed at any time to protrude beyond the biting jaws forming a skin fold. Such flanking folds defeat the purpose of the E.C. Left beyond the two biting edges a skin fold will furnish, so to speak, a back door that will allow blood vessels to continue nourishing the trapped flap with the needed nutrients and the needed gaseous exchange. Such back doors result in delaying, if not preventing, the skin flap from withering away. That is why the presented here-in E.C. has one of its jaws extending into two lateral appendages imbricating into two nicked notches made in the other biting jaw. This arrangement ensures the containment of the skin flap, preventing it from pocketing beyond the biting edges.

Needless to say, the length of the biting edges should be longer than the basal furrow of the bitten skin flap, to allow for its expected lateral expansion upon bringing the biting jaws to a deadlock. Securing a suitable E.C. to a fatty skin flap, will be easier on well-defined skin folds. These could be either due to loss of weight on the part of the patient or could be due to loss of skin elasticity, as taking place in the elderly. However, in younger patients, a turgid skin flap may require kneading by the physician to establish a good holding furrow at the base of the flap for the biting edges of the E.C. to sink in.

The described E.C. in this disclosure has a straight biting edge. Yet E.C.s can come in different sizes and shapes to fit the different skin flaps to be severed. Those with straight edges are suitable for severing the "double chin", (Figure 5A.) They are also suitable for severing the "wobbly grand,ma flap" hanging along the inside of the upper arm, (Figure 5B.) On the other hand, E.C.s with convex edges may be suitable to oust smaller flaps in awkward places. While those with concave edges will be needed to sever the "love handles", and the like, (Figure 5C). Still bigger fatty flaps, like the nicknamed "spare tire" necessitates a different approach.

The medical practitioner may find it suitable to chip it off bit by bit over a longer period of time. Or he may choose to use, e.g., three concave E.G.s fitting side by side around the semicircular fatty flap. In this case, the middle E.C. will have neither flanking appendages nor nicked notches at its biting edges. The arrangement of an appendage imbricating in a nicked notch will only be a peculiarity of the outer most edges of the two E.C.s on either side of the middle one.

Evulsion clamps are intended to do different functions. The first is to shut down the blood supply so that the flap is forced to starve and eventually to snuff out. The second is to force the blood capillaries on the other side of the trapped flap to change course, to anastomose, and to form a continuum with the circulatory system of the patient. The third is to induce the formation of an abscission layer where the squeezed skins are forced to meet. This abscission layer will be forged from the crushed cells. And it will be comparable to the dried pellicles protecting our healing wounds. Eventually the moribund skin flap separates at the abscission layer hopefully favoring a seamless bond. However, if it does not, it can be helped with minimal medical intervention.

Comparable to a healing cut, severing a skin flap is expected to take about 10 days to be snuffed out. However, the time needed to achieve an end result depends on a multitude of factors. These include, the size of the skin flap; the thickness of the underlying adipose tissue; its site of the abdomen; the age, health, and tolerance of the patient. At times in certain cases, it may be needed to accelerate the withering of a trapped skin flap. To accomplish this, certain drugs known to help tissues to morbidate quickly, can be used on the isolated flap under the supervision of the concerned physician.

The idea of using E.C.s to dispose of expendable fatty skin flaps forces itself on the stage of therapeutic treatments for various credible reasons. First, because of its advantage over dieting and its safety over a multitude of blood letting invasive surgeries. Second because other medical practices lend unmistakable support for its utilization. Of these the given-up castration, the commonplace circumcision, and the uncanny restoring of ousted parts in mishaps. Certainly in the medical *modus operandi*, much less ingenuity will be needed to sever a skin flap with an E.C. if compared with the meticulous efforts executed in the nanosurgeries of reattaching cut-off organs.

Admittedly, we will be stepping into uncharted waters, yet we will learn as we go. Needless to say, that this was and still is the only way to push back the frontiers of the unknown and to put new knowledge into use. And may be one day, weight loss by severing persistent fatty skin flaps or sagging skin folds in the elderly by external means as connoted by the evulsion clamps will be as popular as trimming our nails. And together with adopting the habit of this "partial human molting," will come sprawling fancy "Evulsion Parlors" where superfluous "skin-flap-ousting" will be practiced by authorized indoctrinated in the art, "Evulsopractors".

THE PAGES ENCOMPASSING THE DRAWINGS

THE FIRST PRESENTING

SYSTEM 1

THE PREFERRED EMBODIMENT OF THE "EVULSION CLAMPS"

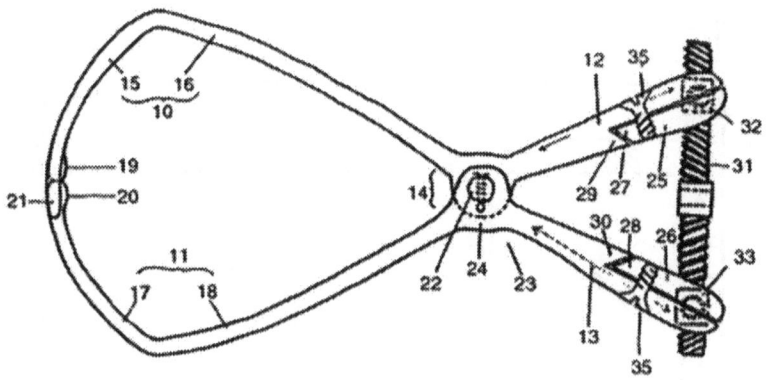

FIGURE I

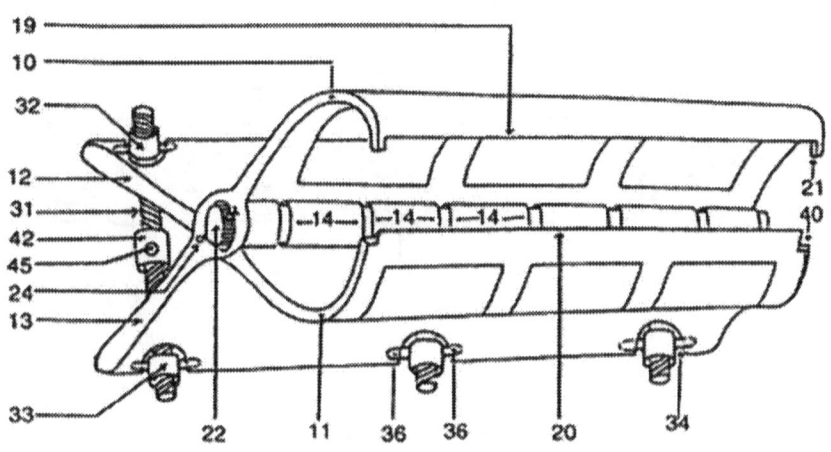

Prospective of the Evulsion Clamp

Wagdy A. Assawah

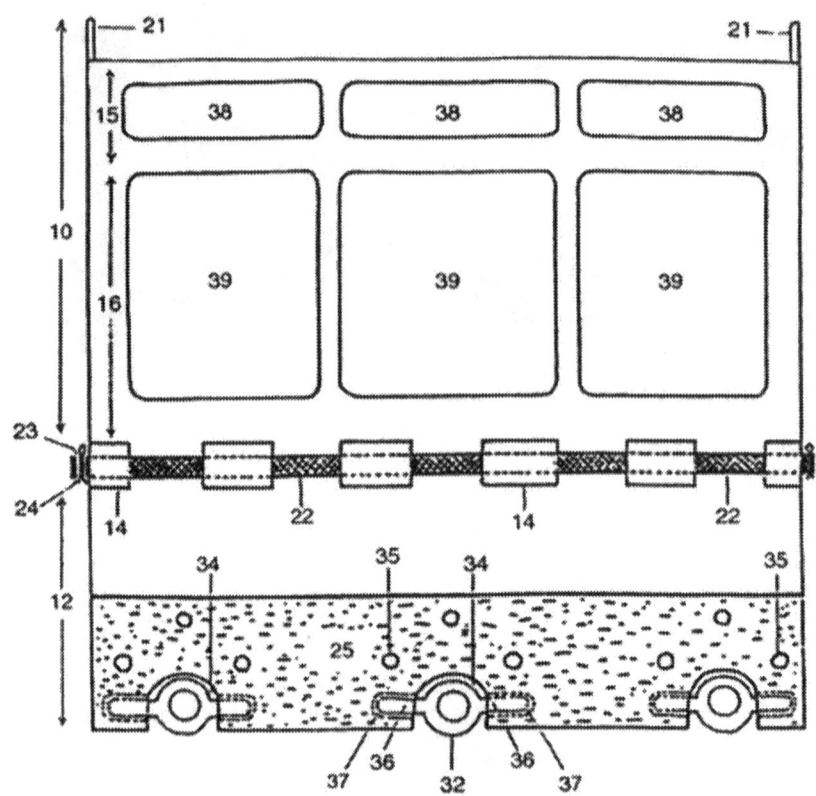

FIGURE 2

The Evulsion Clamp

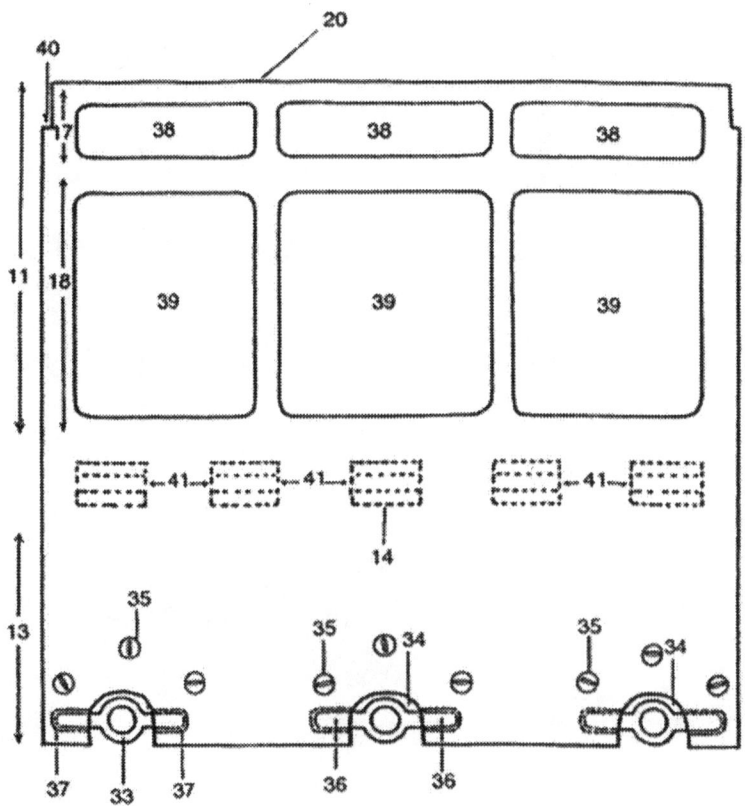

FIGURE 3

The Evulsion Clamp

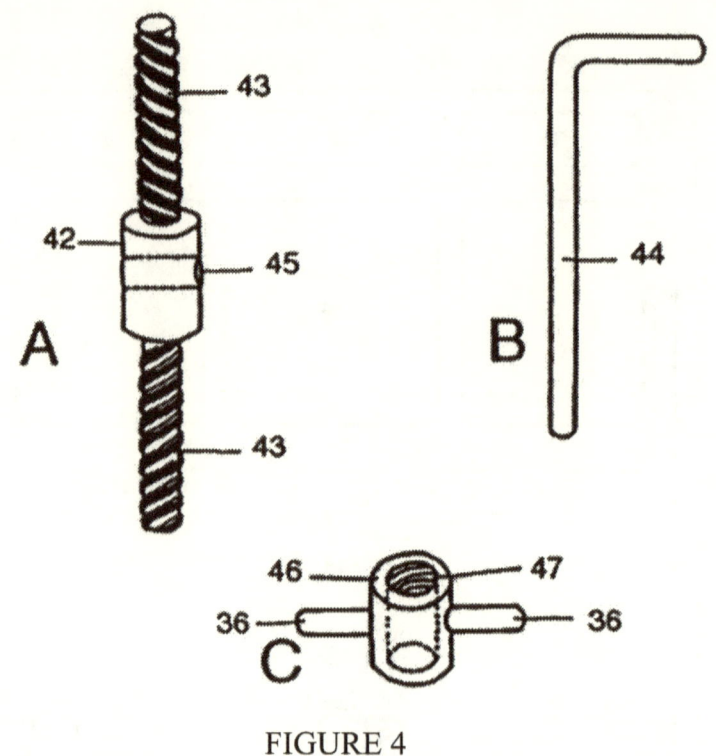

FIGURE 4

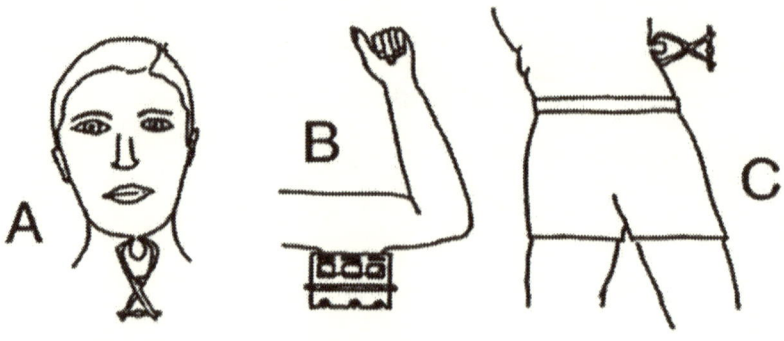

FIGURE 5

The Evulsion Clamp

Addendum:

Coming up with the idea of the "Evulsion Clamp" has a history worth while mentioning. About 60 years ago, I saw in one of my uncle's magazines a picture still engraved in my memory. It was about how the African venus of those times beautifies herself to the men in her tribe. In the picture, the African charmer was inserting in a widened hole made in her lower lip a wooden disc, grooved around its circumference. The bigger the plate-like disc, the more her sex appeal becomes alluring to the celibates in her neighborhood. Other women in the tribe could have a bigger disc for the lower lip and a smaller disc for the upper lip.

Now the question is, was the African man who invented those discs for his female companion aiming at limiting her chatter? That I don't know of! Yet I shall not stop our creative accessory designers from introducing this beneficial concept to our Venuses if so they wish!

Conceited we always down play the ideas of other cultures. On one hand, we praise our contraptions simply because we the westerners or the westernized made them. Take the plates jammed in the lips of the African pixie invented by her paramour, and take the chastity belts invented by husbands for their adored wives. In the first, I see a genuine feeling of love. The African man looked forward to see his sweetheart, the apple of his eyes. In the second, I see the worst intentions. In the name of piety, the European man is displaying his distrust in his wife to the hilt. He made his ugly contraption and went to brag over it. In fact he gave his prisoner motivation to revolt against his domination. No wonder then, when we know that blacksmithing in those days was a lucrative business. The apprentice of this proficiency was thus appreciated because he was the only man who can unlock the cuffs, proffering salvation from frustration.

So much for jesting, let us now go to the serious basics. In a recent TV program about what the anchorman called "The marrying tribe of the Amazon Jungle," the Amazonian man, for some reason, was inserting a short rounded stick into his lower lip. Haunted from an early age by the discs in the lips of African women and now by the short stick wedged in the lower lips of Amazonian men, was to me like the water vapor pushing the lid of James Watt's kettle or like seeing the intruder fungus *Penicillium notatum* killing the growing bacterial colonies in the petri dishes of Alexander Fleming.

Though these lip piercings were used in primitive cultures, yet I saw in them a valuable concept that can be used to rid people from their folded unwanted skin creases. Such creases are usually byproducts of loosing weight. Getting rid of them is a remedy needed by many who will feel healthier when they dispose of them. For that, I surmised a conceptual apparatus that can sever such flaps with considerable ease and with minimal medical supervision. Hence, I took upon myself to design what I called, "The Evulsion Clamp" to do this particular job.

To prove that this invention is not the "Emperor's New Clothes," the concept is strongly supported by other medical practices. In fact, I find a lot of parallelism between my daring suggestion for getting rid of skin flaps and the comparatively laborious technique of lengthening bones in shorter individuals. They both eventuate a disconnection in a continuity to reinstate a new desired perdurability. In the bone lengthening technique, the bones in the legs, i.e., tibiae and fibulae, are deliberately broken in both legs. The broken ends are then kept apart by stainless steel encagements, one for each leg. These encagements called also scaffoldings, besides widening the gaps in smaller increments between the broken ends, also support the patient during his extended treatment. This allows the separated ends to produce new bone growths that help the ends to knit together, resulting in a taller, contented person.

These evulsion clamps, together with my earlier suggested minimal caloric foods will hopefully offer our dieting populace and others around the world, healthier physiques and to be more lighter on their feet. And mark my word, this invention, now looked upon by the puritanicals as an undoubted absurdity, will in time be a ratiocinative marvel, proving that it is unwise to resist unlikely ideas. Once upon a time, a prodigious man whose TradeMark is $E=MC^2$, eloquently said, *"Only daring speculation can lead us further, and not accumulation of facts"*. --**Albert Einstein.**

VI. (B) ON HOW TO EVISCERATE A DANGLING BELLY FLAP

BY

WAGDY A. ASSAWAH

Notarial Seal dated   April 9, 2002

The following is a treatise to supplement my earlier invention entitled:

"Evulsion Clamps To Sever Skin Flaps"

Introduction:

With the completion of the earlier mentioned contribution, I came to the conclusion that the evulsion clamps are best suited for the removal of smaller human fatty skin flaps, but not for the removal of larger ones like those skirting the bellies of portly individuals.

Wagdy A. Assawah

Discontentment in scientists and inventors is a blessing in disguise, pushing the ardent to excel over themselves. Earlier I was dissatisfied with my "Simple Battery" for desalting seawater and consequently contrived the composite, "Bee-hive Battery" that can work around the clock, and expectedly should yield considerable amounts of potable water. Now, dissatisfied with my earlier E.C.s. I decided to build something new, closer to the edge, to improve over my first prototype. Henceforth I came up with a second generation that can handle these cumbersome dangling belly flaps with ease and with more efficacy.

For a fact, the removal of tissue masses by snaring is known in medical practices. So tonsils, polyps, and the like are removed by instruments called the "Snare-boxes". This instrument consists of a wire loope that can be constricted by a mechanism in the handle of this box to get rid of those disturbing tissues. The question was then, why the bigger bothersome fatty flaps dangling from our bodies were not given the deserved attention by those in the medical arena? Was this because of their intimidating sizes? Or was it because, when needed, their removal by surgical means was the first to cross the minds of those skilled in wielding their scalpels with ease?

However, there is logic in discovering another safer non-blood letting alternative for getting rid of those nocuous belly flaps. In the suggested here-in method, new contrivances, i.e., "The Snare-Tackles" are used to eviscerate those unhealthy flaps. And why snare-tackles was coined for a name, that is because one part of the contrivance will be responsible for hoisting and pulling and the other part will be responsible for snaring the tissue mass needed to be eviscerated.

In the earlier submitted manuscript to the P.T.O. dated 03/03/1999, I was committed to justify my idea from a medical point of view. And since I have already done that, I find it plausible not to repeat my justifications once more, but to discuss right away, "The Snare Tackles" from a structural point of view, then from a working point of view. In its action, a snare-tackle, is comparable to a snare-box used in tonsillectomy and in the removal of pedicellate polyps. Yet for an intrinsic difference, while the snare boxes deal only with tissue masses ending on neck-like pedicels, the snare-tackles will be dealing with a tissue mass connected and integrated with the patient's abdomen by a continuum extending from one side of the body to the other.

The Snare-Tackle

From a structural point of view

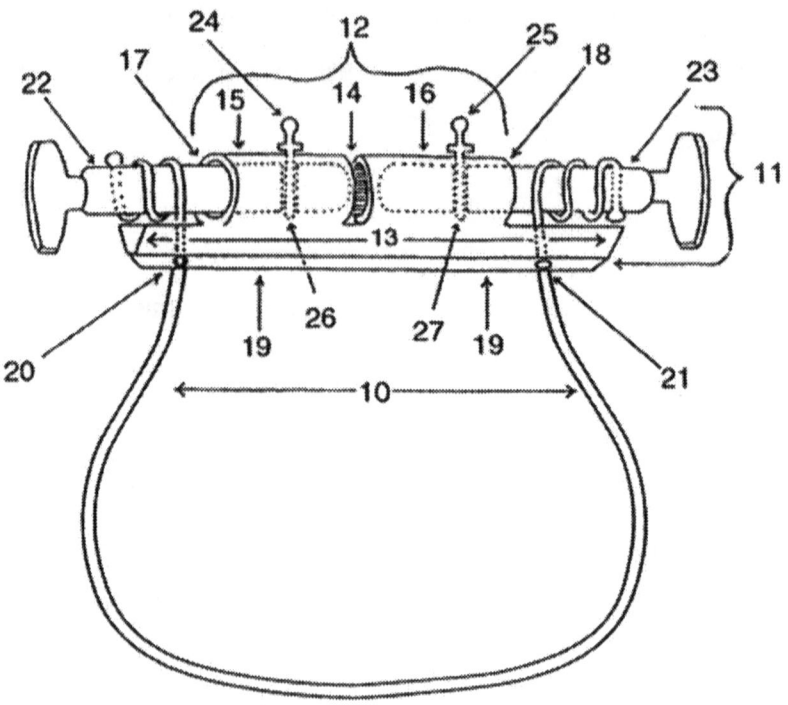

FIG I. Showing a projection of the Snare Tackle.

The Assemblance In Fig. I is Showing:

A loope made of either wire, cat gut, or nylon forming the snare noose (10) and a central piece (11) made of a light metal. This latter comprises an upper cylindrical component (12) and a lower blade component (13). The cylindrical component (12) is divided medially by a septum (14) to eventuate two antipodal tubules (15) and (16) with antithetical mouths (17) and (18). The cylindrical component (12) is integrated with and rests on the blade component (13).

In transverse section, the blade is trapezoidal in shape with the narrower end constituting its straight free edge (19). The blade has two holes (20) and (21), one in each flanking protraction extending beyond the super-imposed cylindrical component (12). The holes (20) and (21) allow the two free ends of the snaring noose (10) to pass through the blade to hook up with the corresponding handed-bars (22) and (23). The free endings of these bars, having handles, will be responsible for hoisting and pulling the two ends of the snaring noose (10). Meanwhile, the continuous endings of the two handed-bars remain housed into the antipodal blinded tubules (15) and (16) respectively.

Due to the possibility of back coiling upon ending the tightening of the snaring noose (10), by stopping the hoisting and pulling made by the handed-bars (22) and (23), pegs (24) and (25) are used to prevent such back coiling from happening. So when needed to maintain a tight grip on the lassoed tissue mass, the pegs (24) and (25) are passed via apical apertures made in the upper walls of the blinded antipodal tubules. From there, the pegs are then inserted into corresponding holes (26) and (27) made in the respective continuous endings of the two handed-bars (22) and (23) to secure them in place after every tightening episode.

THE SNARE TACKLE:

From A Working Point Of View:

Removal of the skirting belly-flaps, necessitated a totally different approach from that used in removing pedicellated tissue lumps. Morphologically belly-flaps usually extend from one hip to the other. Thus, one cannot seize them, so to speak, by the neck to tie a ligature around their narrow basal parts or use a snare-box on them. Accordingly, it is suggested here-in to segment the skirting belly-flap into portions. Then to deal with the portions separately, yet simultaneously. To explain how this is done, i.e., the segmentation, the snaring and eventually the evisceration of the skirting belly flap, the following sequential figures II to VII together with their concomitant texts are forwarded to rationalize this new break-through technology.

This process, which I definitively give the name "Bio-Chiseling" entails getting rid of extra body tissues that are eclipsing a congruity of an exquisite continuity. From the start, this bio-chiseling was goaded by much the same revelation in making the statute of "David". Once Michelangelo was asked, "How did you sculpt David?". His intuitive answer, "I saw the semblance inside the rock, and all that I have to do, was to peel off the layers blanketing it!"

Seeing the weighty dangling flaps discomforting portly people, I knew that I can free them, from the load they are destined to carry around for the rest of their lives, with ease and with minimal medical complications. The needed instruments will not be the gifted chisels of Michelangelo, but an equally effective instrument, "The Snare-Tackles".

Wagdy A. Assawah

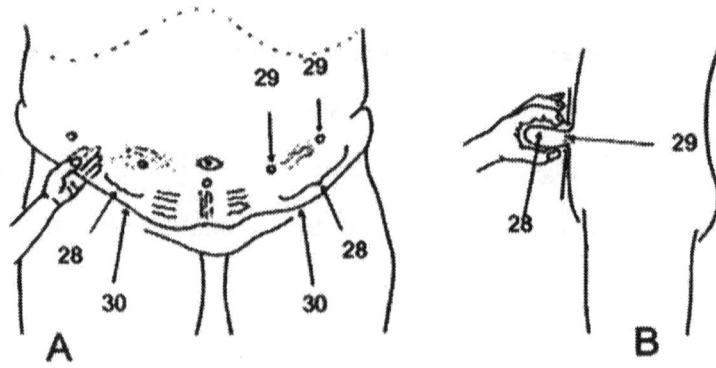

Fig. II

Fig II-A is showing a front view of the skirting belly flap (28) extending from one hip to the other and with its free end (30) dangling from the patient's waist. The under-fold, in advanced cases, can easily accommodate the fingers of the palm while the thumb is left to hold the upper side of the flap. Also, showing the sites of punctures (29) (five in the drawing) made for threading the free ends of the snaring nooses so that they can hook up with the corresponding snare-tackles, that will come to rest on the upper side of the belly flap. In so doing, the punctures represent the first step in the segmentation process of the skirting belly flap (S.B.F.). Subsequently making several smaller portions easier to manipulate for achieving an ambitious end in-sight.

Fig. II-B is showing a side view of the S.B.F. (28), lifted up to show its basal stemming part where the punctures (29) are to be made.

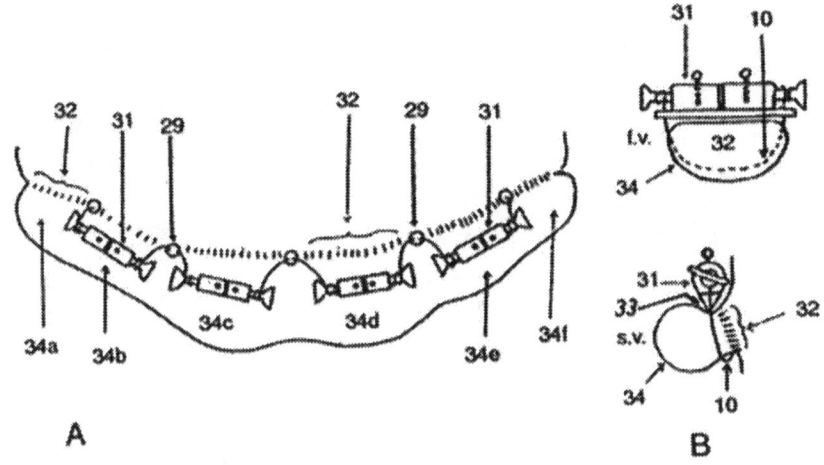

Fig. III

Fig. III-A is showing the mounted snare-tackles (31) in top view and the punctures (29) through which the ends of the wire snaring nooses (10) are passed to hook up with the handed-bars of the super imposed S.T.s (31). The five punctures and the four S.T.s divide the S.B.F. into six portions (34a to 34f). The hatchings (32) demarcate the inner most boundary of the belly flap i.e., the extended basal stemming part amalgamating the flap with the abdomen of the patient.

Fig. III-B is showing a front view (f.v.) together with a corresponding side view (s.v.). Both views show that the wire snaring noose is encircling the basal stemming part (32) of a lassoed portion (34) before working the S.T.s (31) resting atop of it. Also revealing the elbowing (33) of the blade edge (19) into the portion to be eviscerated, yet the portion (34) is still plump and lissome.

351

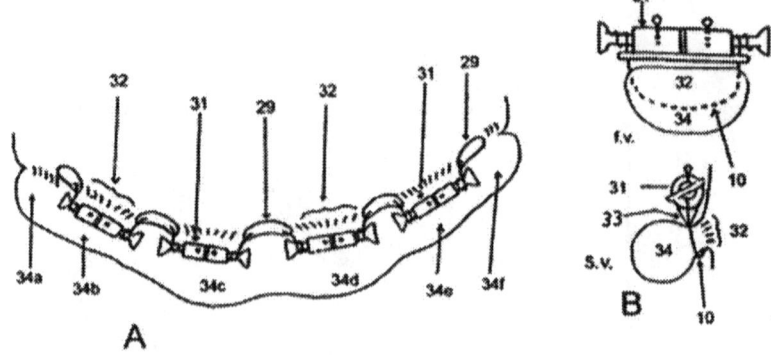

Fig. IV

Fig. IV-A is showing the first stages in tightening the snaring nooses (10). The continued wringing of the basal stemming parts (32) by working the snare-tackles (31), besides widening the narrow punctures (29), they also usher the first stages in diminishing the sizes of the basal stemming parts (32) of the six portions (34a to 34f) connecting the S.B.F. with the patient's abdomen.

Fig. IV-B is showing a front view (f.v.) together with a corresponding side view (s.v.). Both views show that tightening of the snaring noose (10) together with the elbowing (33) made by the blade edge (19), is continuing to diminish the basal stemming parts (32) setting the stage for the complete evisceration of the belly-flap for good.

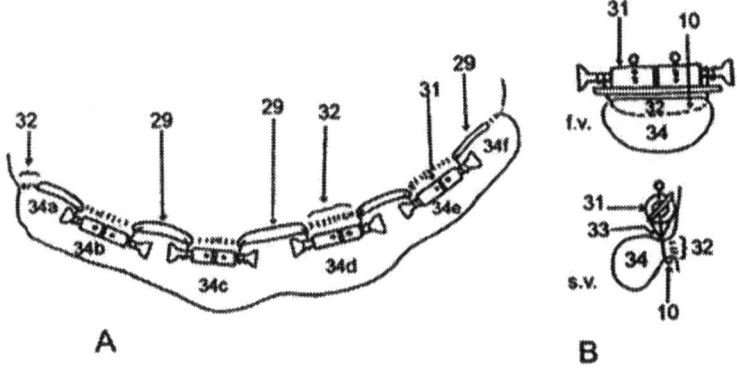

Fig. V

Fig. V-A is showing the effect of continued wringing of the basal stemming parts (32) holding the portions (34a to 34f) comprising the skirting belly flap of the patient. The smothered basal stemming parts are now reduced to narrow nexuses. Also showing that the openings (29) are developing into agaping slits i.e. fenestrae, separating the belly-flap per se, from the abdominal wall, leaving the flap hanging by slimmish nexuses.

Fig. V-B is showing a front view (f.v.) together with a corresponding side view (s.v.). Both views clarify the doings of the snaring nooses (10) worked by the super-imposed S.T.s. on the basal stemming parts (32). Now reduced to vulnerable nexuses, they can be easily incised by ligatures, modified snare-boxes or surgically by "The cutting-stop-bleeding" scalpels, also called "The Cauterizing Hot Knives".

Wagdy A. Assawah

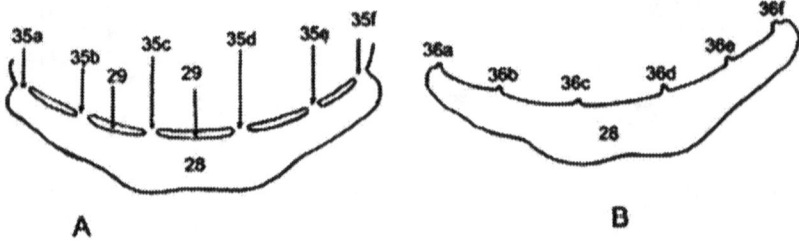

Fig. VI

Fig. VI-A is showing the skirting belly-flap (28) after the removal of the super imposed snare-tackles. The flap is still connected to the patient's belly by six isthmic nexuses (35a to 35f).

Fig. VI-B is showing the belly-flap (28) after being eviscerated from the patient's abdomen. The six isthmic nexuses, that were putting to the last moment the belly flap in conjointment with the patient's body, are now severed leaving behind stubby remnants (36a to 36f) in their places.

P.S. Needless to say, according to the size of the S.B.F., the portions can be increased in number by making more punctures and using more S.T.s, or the portions can be reduced in number by making less punctures and using fewer S.T.s.

P.P.S. Being byproducts, the eviscerated flaps may come in handy in skin grafting operations needed by those harmed in fires or accidents, provided they make perfect matchings.

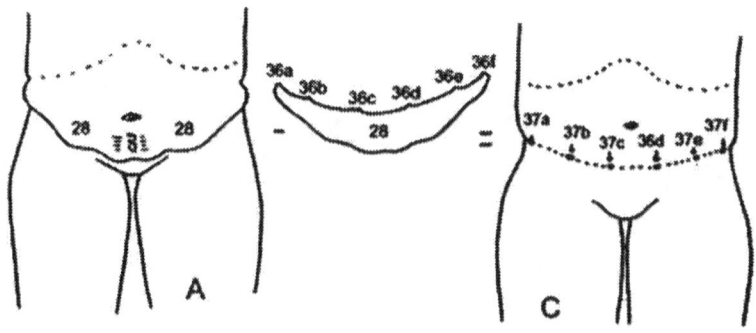

Fig. VII

The Summation Of An Equation
If Not An EUREKA MOMENT, What Is It Then?

Fig. VII-A is showing the cumbersome belly flap (28) dangling from the patient's abdomen.

Fig. VII-B is showing the eviscerated belly flap displaying rudimentary scars (36a to 36f) in place of the done with nexuses that were holding the flap to the patient's abdomen up to the last minute.

Fig. VII-C is showing the pockmarks (37a to 37f) after severing the nexuses to extricate the patient from his scatheful belly flap. Compared to dieting, the process gives the client in a trice what he has been dreaming of i.e., bona fide flattened abs. And this in itself is nothing short of being a miraculous achievement.

Irrefutably let me say, "Throw away your errant cannulae and get to the slow but sure, snare-tackle riskless methodology".

REFERENCE NUMERALS IN DRAWINGS

10.   Snaring noose.

11.   Central piece of snaring-tackle (S.T.)

12.   Cylindrical component of central piece.

13.   Blade component of central piece.

14.   Septum in cylindrical component.

15.   and 16-Two blinded antipodal tubules.

17.   and 18-Two mouths of the two blinded antipodal tubules.

19.   Free end of blade component.

20.   and 21-Two holes in the free end of blade component.

22.   and 23-Two opposing handed-bars.

24.   and 25-Two pegs.

26.   and 27-Two holes for accommodating the two pegs.

28.   The skirting belly flap (S.B.F.).

29.   Punctures widening into fenestrae in the S.B.F.

30.   Free end of the dangling S.B.F.

31.   Snare tackles resting atop the S.B.F.

32.   Hatchings demarcating the basal stemming part of the S.B.F.

33.   Blade edge 19 elbowing into the lassoed portion of the S.B.F.

34.   Six portions of the S.B.F. designated by subscripts a to f.

35.   Six isthmic nexuses designated by subscripts a to f.

36.   Six stubby remnants on the eviscerated S.B.F. designated by subscripts a to f.

37.   Six pockmarks on the patient's abdomen designated by subscripts a to f.

Citing The Benign Biological Consequences

Of The Flapcision Breaking-Through Technology

As a child, you may have avoided the ceremonial rite of circumcision. However, upon passing your youth years, a **Flapcision-rip** will be an embraced extrication rite. Also it will represent a piece of tissue you wouldn't want to reinstate. And here is why.

1.   Doing away with some or all the flaps; namely the belly flap, the right and left flaps of the upper abdomen, the flaps in the back of the upper arms, the groin flaps, the double chin, will for sure give the patient a new "Youthful Posture".

And for an appreciated outcome, he will be overjoyed by being **lighter on his feet.**

2.   Certainly the bulky flaps storing our malign byproducts, besides shortening our lives, are also overworking our skeletal systems, hearts, lungs, kidneys, glands...etc. Now by physically reaching out and selectively biochiseling them, will give the patient hope for a Healthier Existence.

And for a cherished aftermath, he will be euphoric enjoying a **new rejuvenated virility.**

3.   The successive evulsions of, so to speak, the slum areas of our bodies exhausting our catering organs, will help the afflicted to enjoy a happier temperament. One thing for sure, the bio-chiseled individual, will be more "Comfortable In His New Casing."

And for an unbelievable inference, he will be **jubilant for the loss of what he once thought to be a "Life-time Burden".**

Comparable to Prometheus who mythically stole a flame from Mount Olympus, to give it to mankind, the good riddance of the dangling flaps will certainly offer man a longer life span. Consequently, the evisceration technologies will be no less than veritably pilfering a sparge from "The Fountain of Youth".

A Warranted Annotation:

The forwarded methodologies were thought of to relieve the patients from their different fatty skin flaps. Most probably, they will be met with resentments if not by acerbic denial. Comparable to an unnoticed beam of rays coming into a dark room, the presence of light will only be envisaged by the sentient from the reflections made on the particles floating in its path. As to the approbation of both the evulsion clamps and the snare-tackles, their potential will only be grasped by the ratiocinative even before the beneficial ends justify the utilized means. Logistically, for a righteous resolution we are only allowed to judge things by their experimentally proven results and not by arm-chair speculative objections.

For a first encounter with these methodologies, the doubtful will be troubled by their simplicities. And though they may not stand on equal footing with the imposing discovery of the **Double Helix** by Watson and Crick which, via its genetical applications, can improve us on the inside, the forwarded here-in pernickety will do us no less improvement on the outside. And this optimism is born of pantoscopic consultation of different biological disciplines.

Wagdy A. Assawah

In addition to the antecedent, I cannot help thinking, had I chosen to study medicine instead of biology, my overriding concern could have been the welfare of the patients with no time left to think of those irritating fatty skin flaps. Hence I dare say, with my leitmotiv passion for solving problems, I felt decreed to unriddle and consequently end the sufferings caused by those unwelcomed fatty flaps, especially the ones skirting the bellies of the afflicted. Thus, I was intuitively steered to eventuate the previously mentioned gadgetries to end the anguish of the sufferers.

Henceforth, if these daring technologies come about and take hold fulfilling my expectations, I should then take all the credit and eschew all acerbities. And to show that I have no doubt about the probity of my foreseen revelation, I am willing to be the first to go under the snare-tackles to get rid of my own bothersome belly flap. Yet this one of a kind operation, **"The Flapcision Rip"** should be done by noted surgeons since they will be the counselors responsible for laying out the rules and for setting the pace for a new **"Medical Terra Incognita"** to make of it a textbook case. Hopefully this will also establish an unheard of profession that will be practiced by a sector of medical trainees for whom I subjectively choose the name, "The Evalsopractors".

Though some may consider the following as wishful thinking or worst, unfounded vanity, yet my sixth sense tells me that ten years after my great leap in the dark, I'll be depicted on a U.S.A. commerative stamp stating, "Though not a physician, yet he instituted, **The Flapcision Technologies**".

Citing of the above is not meant for pompous bragging. It is only intended to be a testament for: **first** to stress my sturdy belief in the righteousness of the perception; **second** to attest my unwaivered confidence in its far-reaching beneficial outcomes; **third** to assure that ascendancy will only come to those who dare to begin.

Wagdy A. Assawah

ON TRIALS TO PUBLICIZE MY HUMAN-BIO-CHISELING
TECHNOLOGIES
"The Resentment and The Resolve"

Knowing that my methodologies for ousting our skirting belly flaps are going to relief the afflicted of their scatheful burdens that cannot go away by dieting, I decided to publish them with medical journals. Unfortunately they were turned down by *Annals of Internal Medicine* on April 3, 2003 and by *Journal of American Medicine* on April 9, 2003. Also, I never heard from Dr. William Kettyle of MIT-Medical, Cambridge, MA.

Disappointed over their refusal, I came to the conclusion that they may have equated me with tricksters unlawfully practicing medicine in tenebrous clinics. Certainly this envisagement is far from right. Equating impostorism with intuitivism of an avatar of the future who insightedly contrived a useful medical practicability is illogical and wrong. Steadfastly believing that what I came up with, is a genuine panacea capable of winning the battle of the flap, I decided not to take their refusal lying down, *per contra* vowed to fight to the bitter end to bring it to the limelight. Thence, I wrote back explaining that my contrived technologies are literally capable, of reaching out for a dangling flap and bloodlessly severing it for good.

364

Hence offering the afflicted flat abs in a trice, and this is nothing short of a miraculous achievement. Fulfillingly, the introduced are no less than, "Weapons Of Mass Reduction" created to liberate us from our superfluous flaps.

Attempts To Disseminate The Good News:

Defending the uncommon, necessitated sending copies of the work to show its plausibility, to explain the contrived gadgetries and to expose its innate potentialities. The contacted were:

The Family H.M.O. Doctors:

Cardiologist, Mr. Clifford Strauss, D.O.Fac. (Nov. 27/ 2002)

Ophthalmologist, Mr. Barry N. Kutner, M.D. (Dec. 26/ 2002)

Dentist, Mr. Raef Marcus, D.M.D. (Dec. 26/ 2002)

Physician, Mr. Ibrahim Mady, DPH (Dec. 26/ 2002)

Wagdy A. Assawah

The asked question, "Am I a jaunty optimist or tricked by a pipe dream?" It was only Dr. C. Strauss, who was pleased I shared my idea with him.

The Philadelphia Inquirer (Jan. 20/ 2003)

For the *Inquirer,* I reminded them that their journal alerted America and the English-speaking world about Mr. Franklin's "Kite and Key" experiment. Now they can make America aware of another "Philadelphian" and his "Human Bio-chiseling" life changing technologies for the afflicted, to do away bloodlessly with their skin flaps.

The Presidency at Washington D.C. (May 29/ 2003)

For the White House, I pointed out that my suggested technologies for ousting our dangling belly flaps is a future desirability for which time has come. That revelations are only enjoyed by the intuitive and not necessarily by the involved in the art. My pegged plea, "Please consult the versed and don't let these technologies go unquestioned and unexplored."

366

The Medical Reporter of NBC T.V. Station (June 18/ 2003)

I wrote to Ms Cherie Bank inquiring: Is the author suggesting a genuine practicability or a disingenious malarkey? That my first generation of this invention was granted a patent, thence any possibility north of zero should be considered and discussed. Consequently, depriving the afflicted of this benignity because the inventor is an outsider of the profession, is to say the least an odious act.

Medical Schools and Hospitals (June 19/ 2003)

Manuscripts with appropriated covering letters were sent to:

The School of Medicine, Univ. of Cambridge, U.K.
The School of Medicine, Univ. of Nottingham, U.K.

In greater Philadelphia area, the manuscripts were sent to:

Thomas Jefferson Univ. Hospital.

Albert Einstein Medical Center.

Hahneman Univ. Hospital.

Temple Univ. Health Science Center.

College of Medicine, Drexel Univ.

In the covering letter, I requested the Honorable Members on the Medical Boards to see for themselves that the idea is not from a crackpot but from an insighted scientist. Not to dismiss it out of hand without studying it, that is because the idea has big potential and can be a lightning rod used to further a good cause. Also to ask themselves, "Is it a Medical Marvel? Or is it a Witch Doctor's brew?"

With no response from all of the above, I made up my mind to take my case to "The Mother Of All Courts", the Peoples' of America. Whether thumbs up or thumbs down this will be beside the point. What really matters is the coveted truth and this will only be found in the league of the shrewd and not with the puritanicals.

AN UNEXPECTED ENDORSEMENT FOR MEGA-FAT STRIPPINGS

In an article by Dr. Isadore Rosenfeld in "*Parade* Magazine" of September 21, 2003, the good doctor reported that researchers at the University of Pittsburgh, found that extremely obese individuals who undergo gastric bypass surgeries to extricate sizable amounts of fat, in 73% of cases type 2 diabetes was cured.

The above mentioned reveals the essentiality of getting rid of our burdensome fats, the monkey wrench thrown into our working systems. Whether done by cannulae, gastric bypass surgeries, or via the author's newly-suggested bloodless bio-chiseling technologies, they all offer a desired end, justified by different means.

The good riddance of fatty folds *in toto,* nicknamed "The Slum Areas" of the body, will on one hand result in a one and only visible outcome, that is getting the patients lighter on their feet. Likewise such disposals will also result in a multiplicity of unseeable inferences associated with the physiological functionalities of the organs catering for our somas. For a supporting attestation

369

pertaining to such functionalities, let us take what was stated by Dr. Rosenfeld as a pointillistic revealer. The purported showed that disposing of fat resulted in curing type 2 diabetes. To the author's mind, this was because in the untreated obese, the pancreases had to spread thin their hormones and enzymes, i:e insulin and glucagon to include the fat laden adipose tissues. Consequently upon extricating these corpulent fats, the relaxed pancreases, have so to speak, a freer hand to allot more excreta for subduing type 2 diabetes. This beneficial effect did not surprise the author, per contra it stood to his expectations.

Concomitantly, the more plucking of our baleful fatty flaps that cannot go away by dieting, the more adscititious physiological benignities are going to be recognized down the road for other catering organs – i.e., liver, glands, kidneys, lungs…etc. and for other catering systems – i.e., circulatory, skeletal, nervous, lymphatic… etc. – working for the rectitude of our bodies.

The bio-chiseling of our metabolically demanding flaps will not only result in curing type 2 diabetes, but also will eventuate cascading life promoting potentialities. Doing away with our dangling flaps will favor new physiological regimens, affecting not only our pancreases but also other catering organs servicing our bodies. Of the expected benisons, the following may take precedence:

- The enjoyment of relaxed hearts due to eliminating the lengthy labyrinthine blood networks permiating the disposed flaps.

- The enjoyment of responsive kidneys because they are no more overwhelmed by the excreta produced by the metabolic activities that were taking place in the ousted flaps.

- The enjoyment of streamlined breathing due to shunning off the extra amounts of carbon dioxide produced from breaking food down in the previously truncated fat laden creases.

In the author's belief, the above benignities coming in the wake of minimizing our sizes represent the tip of an iceberg that needs examining, evaluating, and collating. Such needed studies besides making us healthier, could add years to our life expectancies.

371

Human bio chiseling technologies capable of transforming us from paunchy to lanky are here to stay, in spite of the hot air of the lobbing few. Their unwelcomed efforts may delay the advent of the contrived by the author for awhile. Yet in view of the fact, that they offer a bloodless bonanza to the portly millions in the U.S.A. and around the world, these technologies are destined to prevail. And to those who cannot lead or follow, I request them to step out of the way and to let the afflicted enjoy what they are entitled to, "The Midas Touch."

Whether "Midas" or "Assawah", this touch will literally reach out to truncate the flaps pocketing the invidious fat to offer the patient in a trice, the dreamt of flat abs and the longed for healthier, leaner, and lighter bodies. Howbeit for an intrinsic difference, the first touch conformably delivers in mythical stories, meanwhile the second touch confirmedly delivers in real life.

INTRODUCING A NEW PERIMETER IN OUR MEDICAL ASSESSMENT METHODOLOGIES

From the earlier mentioned, on the importance of the good riddance of the villainous fat accumulating in our dangling flaps, came to the light the indicative importance of knowing about the ratios of muscle mass versus the fat in our bodies. This ratio will always be a tipster on both the health status of different individuals and a clairvoyance on their life expectancies. (See accompanying table) The forwarded tentatively emphasizes the following:

• Individuals having lower "Muscle/Fat" ratios will therefore suffer from overworking their organs and systems catering for their protracted bodies.

Consequently, such disfavored portly individuals will have to accept predestined shorter life spans together with debilitating health conditions.

• On the other hand, individuals favored by higher "Muscle/Fat" ratios will enjoy relaxed organs and systems. As such, these organs and systems will be able to rectify such bodies to their own good ends.

Consequently such individuals will be favored by longer lifespans together with improved health conditions.

• Dieting late in life is unfortunately of little value, because it cannot erase the ills of obesity done over the years. Such ills will continue to undermine the potentialities of the catering organs and systems from doing good by their corresponding somas.

Conclusively, for the sake of our younger generations – the older, fifty and over are beyond salvation except via bio chiseling methodologies. My exemplary advice to the young obese and their families too, "If you don't want your youngster to pay for the sins of portlyness in their older years, I urge you young and old to start dieting right now, and to keep at it." In so doing, the aware will ensure for themselves a benevolent muscle mass on one hand and a deliverance from the malicious fat on the other hand.

Thereafter, the strict adherent zealots to the regimen, will be going into their cognitive years with almost unimpaired organs and systems that are willing to serve them well and to add years to their life expectancies.

Alas, certainly I have wished for someone to warn me, when I was in my prime about this malfeasance of the stored fat in our bodies. However, I cannot blame anyone, then and now. That is because the concept of the "Muscle/Fat" ratios hasn't cropped up, nor was it deciphered for understanding its notoriety.

For a chronological record, up to this point in time, this inference of mine, concerning the effects resulting from the differences in muscle mass versus fat mass ratios, was a shapeless idea in the womb of the anigmatic. Dawning on me this day of Dec. 01/ 2003, I decided to document it and to share it with my contemporaries of all ages and professions. Hopefully the suggested "Perimeter" will be an eye opener to both patients and medical practitioners. To the obese, it will be a reminder to exert more efforts to get healthier and perhaps add years to his or her life expectancy.

To the medical practitioner, to develop the tell tale curves so the patient can know his where abouts from the "Inversion Point". So if he passes it, he is in a declining episode, but if he doesn't, he knows that he is in a favored situation. Now the hoped for this suggested perimeter, is to find fertile grounds, so it can deliver its correctional bounty to the afflicted.

A Tentative Introspective

On the impact of Muscle/Fat Ratios on our lives

Muscle/Fat Ratios	Functionality	The Generated
1% Fat	The needed thermal protective layer	A must for our bodies.
4% Fat	In the acceptable zone	Enough for a backup.
8% Fat	Sub-healthy zone	A red flag. Dieting advised.
18% Fat	Beware zone	Two red flags. Dieting stressed.
35% Fat	Threatening zone	Could result in 2% to 7% loss in life expectancy – Dieting is a must.
50% to 60% Fat	Unacceptable zone	The patient is jeopardizing his catering organs and bringing their work to a trickle. The end is near. Yet surgical intervention can save the day.
60% Fat or more	Deadly zone	Patient reaching the point of no return, needs a miracle to postpone the end.
My message to the obese of all shapes.		Do you hear me now? If you do, then act on it.

A Recapitulation:

For a known fact, sciences started as individualistic subjects. Yet when these early subjects started to blend, our cognizances flourished beyond our wildest imaginations. So astronomers benefited from physicists, physicists from chemists, chemists from biologists, biologists from taxonomists, pharmacists from botanists, environmentalists from ecologists, forensic specialists from zoologists, agriculturalists and animal husbandry specialists from geneticists, archaeologists from geologists, physicians from engineering…etc. Thenceforth, the outcome of the last mentioned connubiality was the coveted gadgetries filling our hospitals and clinics and the establishment of a new knowledgeability i.e. "The Medical Engineering Endeavor".

To the lucid, the contrived implements, i.e. "The Truncating Clamps" and "The Smothering Snare Tackles", though misleadingly inornate, are none other than useful delivering tools, appropriately belonging to "The Medically Engineered Gadgetries".

The coming together of the minds is what nudges us into the unexplored frontiers to increase our understandabilities. So denying a useful suggestion contrived to win the battle of the flap and to offer the afflicted flat abs, before knowing its potentialities, besides being against the logistical, is also against the unwritten commitment to help the sufferers from a hazardous health condition that can shorten their lives.

Unfortunately, the law which has the power to punish a guilty party for a misdemeanor, is unable to punish those who thwart a useful remedy capable of helping thousands-if not millions-from getting healthier and consequently die younger.

HISTORY OF THE NEGLECTED BY HISTORY

OR

HISTORY HAS A HABIT OF REPEATING ITSELF:

On Suggesting Hyperthermia As A Possibility
For Destructing HIV in Vitro to Cure Patients

"Human Bio-chiseling" technologies were not my first intuition in the medical arena. As early as April 04/ 1990, I wrote to President George Bush about combating HIV by hyperthermia. I pointed out that the procedure could weaken the virus, enabling the body to overcome this life claiming scourge.

The forwarded manuscript was entitled, "Could What I Am Suggesting Be a Cure for AIDS". In the communiqué, I mentioned that since the times of Edward Jenner, pathologists used to immunize our bodies against microbes by using killed or attenuated versions of these microbes, or by using sera containing antibodies. Trapped in this classical approach, our microbiologists never ventured other alternatives, i.e. physical means such as circulating blood of the patients in convoluted tubes suspended in a water-bath at temperatures slightly higher than those maintained by our bodies.

Wagdy A. Assawah

The theorized idea was based upon the fact that human bodies usually fight back the invading mesophilic microbes by raising the body's temperature from 37°C to 42°C. Now, who ever stands the excessive heat of the on going battle, whether body or microbe, will be the winner for survival.

Also circulating blood outside the body is not a new idea. Hemodialysis by "The Kidney Machine" is used to filter urine and other waste products. In leukophoresis, it is possible to isolate the heavier leukocytes from the red blood corpuscles. The blood is then turned to the patient after getting rid of the disease inducing leukocytes. With a detailed description of the suggested apparatus, shown here-in, I wrote to the President, Since this is only a suggestion which needs supporting evidence through experimentations that are beyond my abilities, I requested him to send my idea together with the above mentioned apparatus to the concerned AIDS researchers to examine its capabilities. Consequently, on June 27/ 1990, I got a letter from Ms Judy L. Murphy of the "National Institute Of Allergy and Infectious Diseases", saying, "This is in response to your letter to President George Bush concerning acquired immunodeficiency syndrome". Also adding, "We will forward your letter to NIAID's Division of AIDS for further consideration."

Could What I Am Suggesting Be a Cure for AIDS?

By

M. Wagdy Assawah

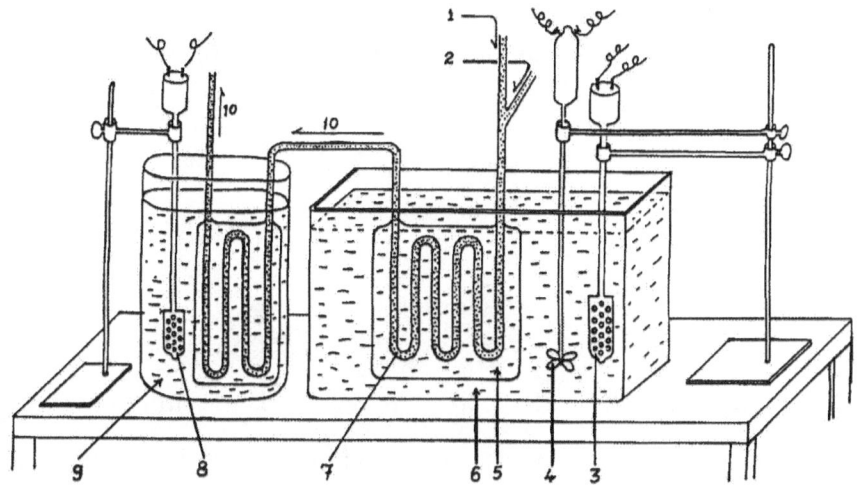

Figure showing a diagrammatic assembly to examine the suggested idea for combating HIV by using suitable temperatures to inconvenience the virus in the blood stream of a patient and at the same time applying drugs harmful to its particles.

Please see accompanying legend explaining the parts of the used apparatus.

Wagdy A. Assawah

Legend of the Contrived Apparatus for Combating HIV

1. Blood from patient.

2. Side tube for adding one of the suitable drugs in a suitable concentration.

3. Temperature regulating thermostat.

4. Stirrer.

5. Plastic pouch.

6. Water bath adjusted to the suitable temperature ($\pm$ 42 °C) which is intended to weaken the virus (HIV) without harming the blood components. The heat could also make the virus (specially when its particles burst out of the T4 cells) more vulnerable to the effect of added drugs, such as AZT, CD_4 or Compound Q.

7. Tubing into which the blood from the patient is passed. The convoluted tubing is held in a plastic pouch (5).

8. Temperature regulating apparatus in a smaller water bath.

9. The smaller water bath is used to cool the blood of the patient to 37 °C before allowing it to return to his blood stream.

10. Arrows showing the directions of the treated blood on its way to the patient.

For a passing remark, this in vitro methodology, though a wild hunch, could open new horizons to subjugate other blood inhabiting microbes. So together with hyperthermia and added antibiotics or other hurtful drugs, we may be able to cure the afflicted from other hazardous pathogens such as:

Rickettsiae causing Rocky Mountain spotted fever.

Bacillus anthracis causing Anthrax.

Clostridium botulinum causing Botulism.

Treponema pallidium causing Syphilis.

Plasmodium malariae causing Malaria.

Trepanosoma gambiensis causing African sleeping sickness.

Now the incapacitated clumps of pathogenic micro-organisms are then removed either by filteration or by differential centrifugation before returning the cleansed blood to the patient. However, we will never know the outcome till we carry out the experiments, get the results, and arrive at conclusions. The hoped for knowledgeability will only come to those who dare to begin. And for those unfamiliar with research, a negative result is also a result worth knowing of. As such it will be the blind alley, to which we don't have to go anymore.

Hearing nothing and not knowing about the fate of my contrived idea, I kept an eye on the news. Eventually, information came to my attention from an article in the *Philadelphia Inquirer* dated Aug 15/ 1990, written by Mr. Robert Byrd. It was entitled, "Doctor Says blood-heating treatment dead." In his account, Dr. William Logan admitted bluntly, his research is almost certainly finished because of a scathing report by federal investigators. However, a partner of him, Dr Kenneth Alonso, went on treating two AIDS patients with hyperthermia. The first, Carl Crawford 33, disclosed the treatment to reporters, pointing to visible improvements and pronouncing himself cured of the deadly and so far incurable disease.

A second patient, who kept his identity confidential, was largely unchanged. Thenceforth, Dr. Logan and Dr. Alonso decided to dissolve their collaboration, yet Dr. Alonso chose to go to Mexico to continue his work.

Now the questions I keep asking myself:

- Why was Dr Logan handed a scathing report?
- For what reason did The Federal Investigators want to abort a promising idea before exhausting its potentialities?

Certainly knowing the reality about this probative methodology shouldn't have been occulted from the people.

In the scientific endeavor, it is an antiquated moral taboo to put locks and keys on free thinking or revelations. That is because knowledge is limited, while intuitive thinking pushes our frontiers, abutting the unknown. Yet the worst thing in our contemporary scientific protocol is the new fad of disengaging ourselves from each other and establishing marked occupational territories to safeguard our interests. Nevertheless, we only benefit when we crisscross our ideas, because in the scientific endeavor we constantly keep correcting our findings to get nearer to the consummate.

Wagdy A. Assawah

VII. A ROTOR DINGHY FOR SHOOTING RAPIDS

Filing Date   07/26/99   Application No.,09/360,919

Foreign Filing License   Granted   08/17/99

A Concise Recountal:

A new rotor-dinghy inaugurates a novelty equipped with an assortment of gadgetries to survive the perils of rapids. It has a bi-lobed bow to resist plunges and a trailing tail for keeping it on a decided course. Most importantly, it is provided with a revolving wheel hugging the lower part of the rotor-dinghy. The revolving wheel is shored with several inflated bolsters designated to protect the craft from collisions with riverside rocks thence to backspin sparing the craft from a counter spin.

Further more, the shored inflated bolsters are designed to spread out to cover a wider circumference to ease a dangerous descent and to autofold around the revolving wheel to allow the water inundating the bolsters to run off with ease, facilitating a gentle ascent. These two movements i.e. the fanning out upon an unexpected dunk and the autofolding around the revolving wheel upon heaving up, mimics the undulating movements made by the mantles of jellyfish. Hence, incorporating these skirting bolsters was with the purpose of emulating Mother Nature's proven methodology.

Wagdy A. Assawah

PATENT APPLICATION

OF

WAGDY A. ASSAWAH

FOR

A ROTOR-DINGHY FOR SHOOTING RAPIDS

Background Field Of Invention:

A rotor-dinghy (Fig. I) was designed with the purpose of over coming the perils of rapids. The contrivance is composed of three parts, An inflated bi-lobed bow for a first part, A rounded body made of sturdy aluminum, referred to as basin for a second part, and A trailing tail for a third part, made of rubber and acting as a rudder. The inflated bilobed bow is fixed to the basin to designate its fore, while the trailing tail is fixed antithetically to the bilobed bow to designate its aft.

The basin is the main part of this forwarded water craft. It is differentiated into an upper part and a lower bottom part. The upper part is cylindrical and caters for both the bilobed bow and the trailing tail. As to the lower part, it has a slanting wall and a flattened underside. The slanting wall is hugged by a similarly slanting revolving-wheel shored with several inflated bolsters. And, though the revolving-wheel together with its bolsters make an integral part with the lower part of basin, yet thanks to its method of mounting, this revolving-wheel can gyrate either clockwise or anticlockwise without hindrance. As to the flattened underside, it is lined with several inflated truncated cushions. The multiplicity of either the bolsters or the cushions is deemed necessary so as when one element gets busted, the rest can still carry out the job, i.e., help in the buoyancy of the dinghy.

Navigators sailing the rotor-dinghy will have to sit on built-in swivel seats. The rounded tops of these seats are provided with foramens, usually four per rim. Crisscrossing their long safety belts and threading them through the foramens, the navigators can anchor themselves to their seats. Due to the endowed ability of the seats to rotate on their supporting columns, the anchored navigators will in their turn be able to respond diligently to the change in course as called for by the twists and turns of the river.

Wagdy A. Assawah

Background Prior Art:

Armed only with their oars, rapids oficionados used to brave the gushy gulches in flat bottom inflatables. At one time, a trial was made using an upright huge wheel to overcome the perils of rapids. The part of wheel coming in contact with the challenged surge has an open vastitude to accommodate the navigators and their equipments. Thus heavier, it was thought to keep the wheel in an upright position. Being under the mercy of fiery swirls, swift currents, and blowing winds, the upright wheel ended up lying on its side. Because mission unaccomplished, its failure resulted in aborting the trial and abandoning the means.

Objectives And Advantages:

The here-in presented art i.e. the rotor-dinghy has several advantages over the conventional flat-bottom inflatables. Endowed with several innovations the rotor-dinghy will neither be intimidated by the raging torrents nor inconvenienced by rocks lining the riverside or differences in water levels.

The limitations of the flat-bottom inflatables can be realized by persons familiar in the art. They lack needed safety measures to counter act colliding with rocks or other floating objects. They are also defenseless against whirls and against sudden dunks when moving from a higher water level to a lower water level. Hence it was a primary objective to provide the rotor-dinghy with several devices for safety, stability, and for keeping it on a decided course. With these convictions in mind, a bilobed bow, a revolving-wheel shored with bolsters, underside cushions, swivel seats, and last but not least a rudder were imperative.

Incorporating a frontal bilobed bow was with the objective of helping the rotor-dinghy to resist head-on plunges and to successfully survive ebbing waters. Being dual in nature, the bi-lobed bow has, two lips that act as a one way valve. Upon plunging headlong, the two lips clamp up, letting the bow to use its turgidity for resisting the plunge. But upon heaving, the two lips disengage to let the scooped water on top of the bilobed bow to ebb easily thus extricating the rotor-dinghy form an under thrust.

Furthermore, incorporating a revolving-wheel shored with inflated bolsters was with the objective of outfitting the rotor-dinghy with several valuable assets. The bolsters act as stabilizers. They together with the underside cushions, help in buoyancy. And both act as shock absorbers when the craft collides with rocks and other floating objects. Also, upon colliding with the riverside rocks or getting caught in a whirler, the revolving-wheel together with its bolsters, backspins. In so doing, they act as brakes, curtailing the expected counterspin of the basin. What is more the bolsters skirting the basin will resist the sudden dunks, by spreading out to cover a wider circumference when the craft falls from a higher water level to a lower water level.

Still, fitting the rotor-dinghy with certain appurtenances, was with the objective to facilitate handling of the rotor-dinghy and keeping it on a decided course. Thus, taking advantage of the rotatability of the swivel-seats, the navigators can respond to the twists and turns called for by the river. Also, incorporating a trailing tail helps in preventing the basin from getting caught in the counterspin caused by the backspins of the revolving-wheel, and enforces keeping the craft on a decided course. This action of the trailing tail emulates the action of the smaller vertical propellers installed at the tapering ends of older copters. While the first curtails the counterspin in rotor-dinghies, the latter prevents the fuselage of the copter from the counter rotation spurred by the horizontal uplifting propeller.

DESCRIPTION OF DRAWING FIGURES

The invention will be better understood when consideration is given to the following detail descriptions there-of. Such descriptions make reference to the annexed drawings wherein.

Fig. I — Showing a medial longitudinal section in the rotor dinghy.

Fig. II — Showing the assembled rudder and components for securing it to the basin.

Fig. III — Showing the internal structure of the inflated bilobed bow using phantasmal vertical sections, namely 44, 45, 46, and 47.

Fig. IV — Showing the rotor-dinghy in top view.

Fig. V — Showing the rotor-dinghy in underside view.

Fig. VI — Showing consecutive stages in assembling the rotor-dinghy.

REFERENCE NUMERALS IN DRAWINGS

11.   Inflated bilobed bow.

12.   Upper part of basin.

13.   Bottom part of basin.

14.   Revolving wheel.

15.   Swivel seat.

16.   Rudder's handle.

17.   Opening admitting rudder's handle.

18.   Upper helm holder.

19.   Upper verge of helm.

20.   Helm.

21.   Proximal end of rudder's tail.

22.   Distal end of rudder's tail.

23.   Lower verge of helm.

24.   Lower helm holder.

25.   Annular protuberance emanating from the lower end of upper part of basin.

26.   Inflated bolster.

27.   Tubby flap arising from lower end of bolster.

28.   Inflated truncated cushions lining the flattened underside of the rotor-dinghy.

29.   Inner smaller cincture.

30.   Inner thickened edge of the truncated cushion pinned down by cincture 29.

31.   Outer bigger cincture.

32. Outer thickened edge of truncated cushion pinned down by cincture 31.

33. Z-shaped annular manacle riveted to the lower part of basin.

34. Lower girder pining down the lower thickened edge of bolster.

35. Upper girder pining down the upper thickened edge of bolster.

36. Lower girder pining down the lower thickened edge of bow.

37. Upper girder pining down the upper thickened edge of bow.

38. Orifice to accommodate the upper verge of helm.

39. Orifice to accommodate the lower verge of helm.

40. The U-shaped part of helm for sleeving the fixed end of the rudder's tail.

41. The tuning-fork end of the rudder's handle to clasp the helm.

42. L.S. in the proximal end of the rudder's tail rigidified by inflated interconnected parallel tubes.

43. L.S. in the tubeless pliant distal end of the rudder's tail.

44. The juxtaposed basal ends of the bilobed bow and their corresponding phantasmal vertical section, (P.V.S.).

45. The bulky middle part, the largest moiety of the bilobed bow, and its P.V.S.

46. The penultimate end of the bilobed bow and its P.V.S.

47. The tip end of the bilobed bow and its P.V.S.

48. An upper concavity shared by the two lobes of the bow.

49. A lengthwise dorsal slit in the upper concavity separating the two lobes of bow.

50. A gape between the two adjacent lobes of the bow extending from the dorsal slit to the ventral dual keel of bow.

51.   The ventral dual keel, displaying two prominent lips.

52.   The penultimate end of the billowed bow starting to zip up to close the gape.

53.   The tip end of the bilobed bow completely conjoined.

54.   Foramens to thread the safety belts of navigators, shown in Fig. IV.

55.   Stout columns stemming from the basin floor.

56.   Smooth rounded heads terminating the free ends of the supporting columns.

57   Tacks articulating the swivel-seats' rotative tops to the rounded heads of the supporting columns.

58.   Buoyed rims forming sockets made in the underside of the swivel-seats.

59.   Annular valluculae in penultimate sites to the free ends of supporting columns.

DESCRIPTION OF THE PREFERRED EMBODIMENT

With reference now to the drawings and in particular Fig. I, thereof the preferred embodiment of the new devise embodying the principles and concepts of the present invention and generally designated by reference System 1 will be described.

Concepts Of System 1:

Specifically, it will be noted that the devise relates to a new and improved water craft for shooting the rapids. The medial longitudinal section in Fig. 1 elucidate the manner in which the different components of this devise i.e. the rotor dinghy, are related.

In its broadened context the devise consists of a rounded body called the basin, made of sturdy aluminum. Its upper part (12) is cylindrical with a bilobed bow (11) indicating its fore and a trailing tail (22) acting as a rudder fixed antithetically to the bow to indicate its aft. The basin's lower part (13) is slanted and perched on a similarly slanted revolving-wheel (14). This revolving wheel

Wagdy A. Assawah

can gyrate clockwise or anticlockwise and is shored with several (six in the drawings) inflated bolsters (26). Meanwhile the flattened underside of the basin is lined with several (six in the drawings) truncated cushions (28). On the inside, the basin is provided with a suitable number (four in the drawings) of swivel-seats (15) that can rotate freely without dislodging from their central columnar supports (55).

Complementary Components For System 1:

System 1 presented here-in is complemented by various other figures, namely II, III, IV, and V to disclose other components, to explain their capabilities and how they perform.

Apropos Of Rudder:

Fig. II is showing the different pits and pieces that make up the rudder. Its trailing tail is made of rubber and is differentiated in two parts, a proximal end (21) and a distal end (22). The proximal end (21) is rigidified by incorporating inflated laterally connected longitudinal tubes as shown by T.S. (42). On the other hand, the distal end (22) is kept tubeless and pliant as shown by T.S. (43). For attaching the trailing tail to its sleeving helm (20), its narrower end is pushed in the narrow space of the U-shaped part (40) of the helm and then riveted. The helm (20) in its turn is also meant to be clasped by a tuning fork (41), terminating a rod-shaped rudder handle (16).

For getting the rudder handle (16) inside the rotor-dinghy, an opening (17) is made in the upper part of the basin. Now with the assembled rudder in one piece, it need to be annexed to the basin. For that purpose, an upper helm holder (18) and a lower helm holder (24) are employed. Each helm holder has an orifice. Orifice (38) in the upper helm holder (18) accommodates the upper helm verge (19). Orifice (39) in the lower helm holder (24) accommodates the lower helm verge (23). With the helm holders fixed in place, the rudder will be ready to help in maintaining the rotor-dinghy on a decided course.

Apropos Of Bilobed Bow:

Fig. III is showing the internal structure of the inflated bilobed bow using phantasmal vertical sections P.V.S. namely 44, 45, 46, & 47.

The bilobed bow has a square, thickened base fixed to the upper part of the rotor-dinghy. The four thickened sides of the bilobed base are pinned down by an upper girder (37) a lower girder (36), and a right and a left girders not shown here due to the nature of the offered representation. Being dual in nature, the bilobed bow shares an upper dorsal concavity (48) and has for a ventral keel, two juxtaposed lips (51) that act as a one way valve.

The first P.V.S. (44) is showing the basal part of the two lobes in juxtaposition. And that they share an upper concavity (48) and share a ventral keel of two primordial lips. The upper concavity (48) displays delimiting prominent ridges and a medial dorsal slit (49) extending the whole length of the bilobed bow from the fixed base to the fused tip (53). In the second P.V.S. (45), the two lobes are comparatively separated, yet still embracing the well developed upper concavity (48). This part of the bilobed bow is the bulkiest part and its concavity displays a wider slit (49) leading to a wider gape (50) which in turn opens to the outside by two thickened lips (51). The third P.V.S. (46) is showing that the two lobes are starting to coalesce at the keel, leaving an upper vestigial gape (50). Meanwhile, the P.V.S. (47) made in the fused tip (53) of the bilobed bow is showing complete fusion of the two lobes leading to complete disappearance of gape 50.

Incorporating a frontal bilobed bow (11) was with the objective of helping the rotor-dinghy to endure the head-on plunges and to successfully survive the under thrusts. Hence, the two juxtaposed lips of the keel clamp up upon pitching headlong letting the bow to use its turgidity to resist the dips. Also being capable of disengaging when buoying, the two lips will allow the waters lodging on the dorsal concavity (48) to sink through the lengthwise slit (49), then to percolate into the gape (50). From there, it passes to the river through the two disengaging lips. Thus, contrary to the belief that loose lips sink ships, these two disengaging lips will extricate the rotor-dinghy from an underthrust.

Apropos Of Revolving-Wheel:

One of the most important parts of the rotor-dinghy is the revolving-wheel and its integrated bolsters. Fig. I is showing that the slanted revolving-wheel (14) is hugging the similarly slanted lower part of basin (13). The revolving-wheel and shored bolsters (26) can freely gyrate clockwise or anticlockwise. This is because it is mounted in such a way to keep it both in place and at the same time allowing it to gyrate freely *in situ*. To achieve this, the upper rim of the revolving-wheel is held in place by an annular protuberance (25) emanating from the lower end of the upper part of basin (12). While its lower rim is held in place by a more or less z-shaped annular manacle (33) riveted to the bottom part of basin (13).

Each inflated bolster has two thickened edges flanking its base. The lower thickened edge is pinned down to the revolving-wheel by a lower longitudinal girder (34). While the upper thickened edge is pinned down to the revolving-wheel by an upper longitudinal girder (35). From each bolster's free end arises a tubby flap (27) that extends the whole length of the inflated bolster.

Seen in top view Fig. IV, the rotor-dinghy is showing that the revolving-wheel accommodates six separate bolsters. Seen in underside view Fig. V, the rotor-dinghy is showing the lower thickened edges of the bolsters are pinned down to the revolving-wheel (14) by lower longitudinal girders (34). Also showing is that each bolster (26) has an extending tubby flap (27) arising from its free end. The underside view reveals also the scabrous surface of the inflated bolster meant to endure the expected collisions with riverside rock and other floating objects. The mounting of these bolsters makes replacing a damaged one with an undamaged bolster an easy task.

Incorporating a revolving-wheel shored with inflated bolsters was with the objective of outfitting the rotor-dinghy with several valuable features. The bolsters act as stabilizers. They help in buoyancy. They act as shock absorbers upon colliding with the riverside rocks or other floating objects. Responding to collisions, the revolving-wheel and its bolsters backspins, preventing the basin from getting caught into an opposite spin, asserting the stability of the basin. This stability is also enforced by wielding the rudder thus keeping the rotor-dinghy on a decided course.

Still yet another important part played by the bolsters is that they resist the sudden dunks when the craft proceed from a higher water level to a lower water level. Forced to plunge the bolsters together with their tubby flaps will fan out parachute style to cover a wider circumference. In so doing, the rotor-dinghy will be able to resist a sudden dip. Counter acting the dip, the rotor-dinghy starts to heave up. This heaving up is facilitated by the autofolding of the bolsters and their tubby flaps thus assisting the waters inundating them to run-off enabling the rotor-dinghy to roll out smoothly. These two movements i.e. fanning out upon an unexpected dunk and autofolding around the revolving-wheel upon heaving up as made by the inflated bolsters and their extended tubby flaps, mimics the undulating movements made in the mantles of jellyfish. These oceangoing creatures are able to ascend, descend, or to pause in place by their synchronized undulations made in their mantles. Hence, incorporating these skirting bolsters was with the aim of emulating mother nature's methodology. Thus adopted, the rotor-dinghies will be rapid worthy crafts and will be safer for the rapids aficionados seeking the euphoria of adrenalin rush.

Apropos Of Swivel-Seats:

Fig. I is showing that the basin is provided with swivel-seats (15). There are four of these seats in the rotor dinghy presented here-in. The swivel-seats have rotative tops mounted on stout columns (55) stemming from the basin floor. The free extremities of these columns form smooth rounded heads (56). These heads end up in centralized sockets made in the underside of the rotative tops. Articulating the rotative tops with the rounded heads (56) of the supporting columns (55) is achieved by the use of tacks (57). Each socket has five equidistantly placed tacks passing through its buoyed rim (58). Meanwhile the protruding five tips of the five tacks embed into an annular vallucula (59) made in a penultimate site from the rounded head (56). This arrangement allows the rotatability of the tops without dislodging from their supporting columns.

Moreover, the rotative tops of the swivel-seats are also provided with peripheral foramens (54). Four for each rotative top. These foramens are for accommodating the safety belts of the navigators. Crisscrossing their belts and threading them in the foramens, the navigators can anchor themselves to the swivel-seats and at the same time can row to safety. Needless to say, the navigators have also to practice how to quickly untie their safety belts when an abandon ship ensues.

Incorporating swivel-seats in the rotor dinghy was done with the objective of securing the safety of the navigators and emphasizing better maneuverability for the craft. Anchored to their seats, the navigators can safely respond to the twists and turns called for by the raging torrents. Furthermore, the rudder-man, taking advantage of the rotatability of his swivel-seat, can maneuver his craft on a safer course. Handling the trailing tail by the rudder-man, is in a sense analogous in action with the smaller vertical propeller operating at the tail end of older copters. In case of the rotor-dinghy, the wielding of the trailing tail helps in negating the antispins caused by the backspinning of the revolving-wheel. Comparatively in old copters, the vertical propeller is installed to prevent the fuselage from getting caught in rotations spurred by the powerful uplifting propellers.

Apropos Of Inflated Truncated Cushions:

Fig. I is showing that the underside of the basin is lined with inflated truncated cushions (28). Meanwhile, Fig. V shows that the truncated cushions are six in number, held together side by side, forming a doughnut configuration. And while their broader bases occupy the circumference, the truncated narrower ends encircle the cleared center on the underside of the basin. This arrangement allows to integrate the truncated cushion to the underside of the rotor-dinghy. The base of each truncated cushion has on its broader end an outer thickened edge (32) and on its narrower end an inner thickened edge (30). The outer thickened edges (32) of the six cushions are pinned down by an outer bigger cincture (31), while the inner thickened edges (30) of the six cushions are pinned down by an inner smaller cincture (29).

Incorporating the inflated truncated cushion to the underside of the rotor-dinghy was for achieving two purposes. These cushions, together with the bolsters, help in buoyancy. And while the bolsters are meant to absorb the impacts with the riverside rocks, the truncated cushions are meant to protect and absorb the impacts with projecting rocks from the riverbed. The ease with which truncated cushions can be installed does not only allow to replace the damaged cushion, but also to protect the rotor dinghy from impairments hence expensive repairs.

Wagdy A. Assawah

ASSEMBLING THE ROTOR-DINGHY:

Fig. VI is showing the consecutive stages in rigging System 1 with its different complementary components.

Assembling the pits and pieces of the rudder displayed in Fig. II was given under apropos of this component. With the rudder in one piece, its handle is then slipped inside the basin through the opening (17). Holding the rudder's helm (20) in place, the rudder proper installment, is thence achieved by securing the helm to the upper part of the basin (12) between the two helm holders (18 and 24).

Concerning the bilobed bow, the four thickened edges of its square base are pinned down by girders hemming all four sides of its base. While an upper girder (37) and a lower girder (36) are shown in Fig. III, the girders on both sides of the thickened square base cannot be represented in a centralized V.S. With both the rudder proper and the bilobed bow in place, the basin is then turned upside down to allow for annexing the rest of the complementary components of the rotor-dinghy.

The first to fall in place on the inverted basin and to rest on its slanted bottom part (13), will be the similarly slanted revolving-wheel (14) together with its shored bolsters (26). The wider end of the slanted revolving-wheel is then allowed to dwell under the annular protuberance (25), emanating from the lower end of the upper part of basin (12). To hold the revolving-wheel in place a more or less z-shaped annular manacle (33) is lowered in place and riveted to the lower part of basin. Though held in place by both the protuberance (25) and the manacle (33), the revolving-wheel leaves a small gap separating it from the hugged lower part of basin (13). This loose fitting of the revolving-wheel together with performing its task in an aqueous medium, allows the revolving-wheel to dextrogyrate or to laevogyrate freely.

The last component to be conjoined to the rotor-dinghy will be the inflated truncated cushions (28). They are juxtaposed on the underside of the basin to form a doughnut configuration. Pinning them down to the bottom of the rotor-dinghy will be accomplished by the use of an outer bigger cincture (31) and an inner smaller cincture (29). The outer one will be for hemming the outer thickened edges (32) of the truncated cushions, while the inner one will be for hemming the inner thickened edges (30). Thus assembled the rigged rotor-dinghy will be ready to challenge the perils of rapids and to persevere.

While the above description contains many specifications, these should not be construed as limitations on scope of the invention. Other variations in size and shape are possible. As for size, smaller or bigger crafts can be built to fit a particular purpose. Concerning shape, longer twin rotor dinghys can be built. These are made by installing two revolving-wheels together with their shored bolsters to surround two separately protruding rounded bottoms coming out of the elongated underside of the twin rotor-dinghy basin.

PAGES ENCOMPASSING THE DRAWINGS

WITH THE FIRST PRESENTING

SYSTEM 1

AS THE PREFERRED EMBODIMENT OF THE

"ROTOR-DINGHY"

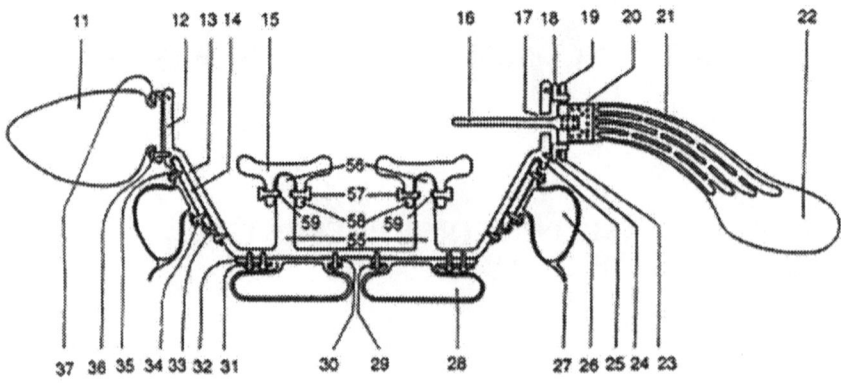

Fig. I Showing a medial longitudinal section in the Rotor-dinghy

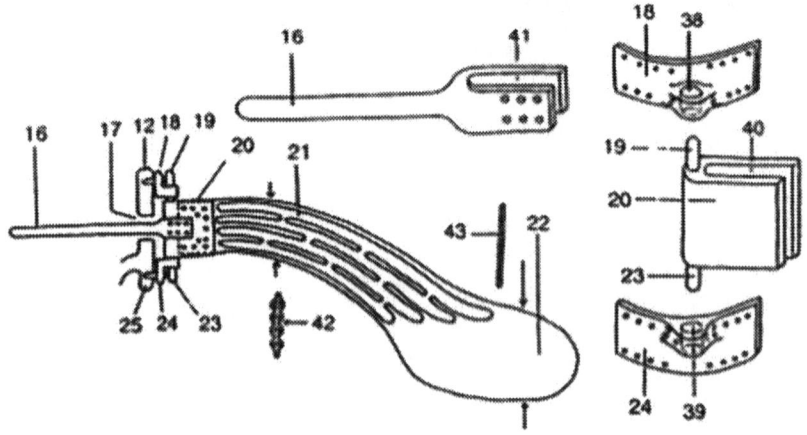

Fig. II Showing the assembled rudder

and components for securing it to the basin.

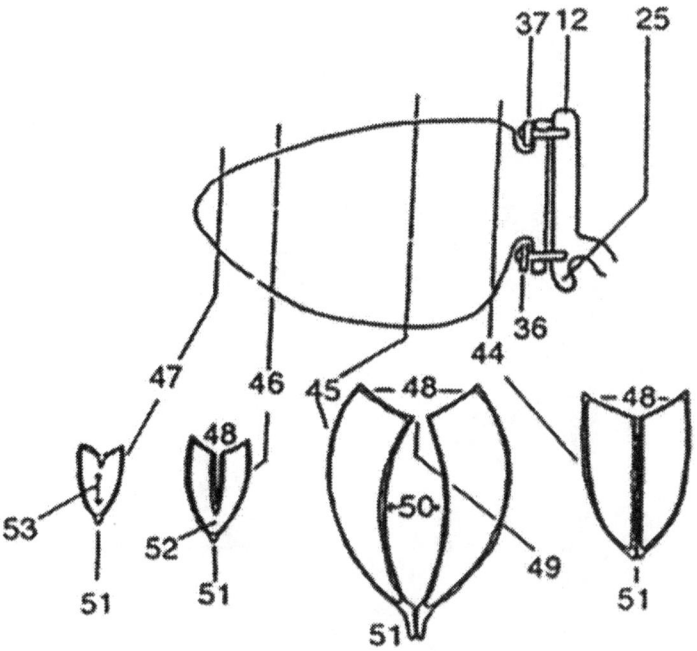

Fig. III Showing the internal structure of the inflated bilobed
bow using phantasmal vertical sections, namely 44, 45, 46, and 47.

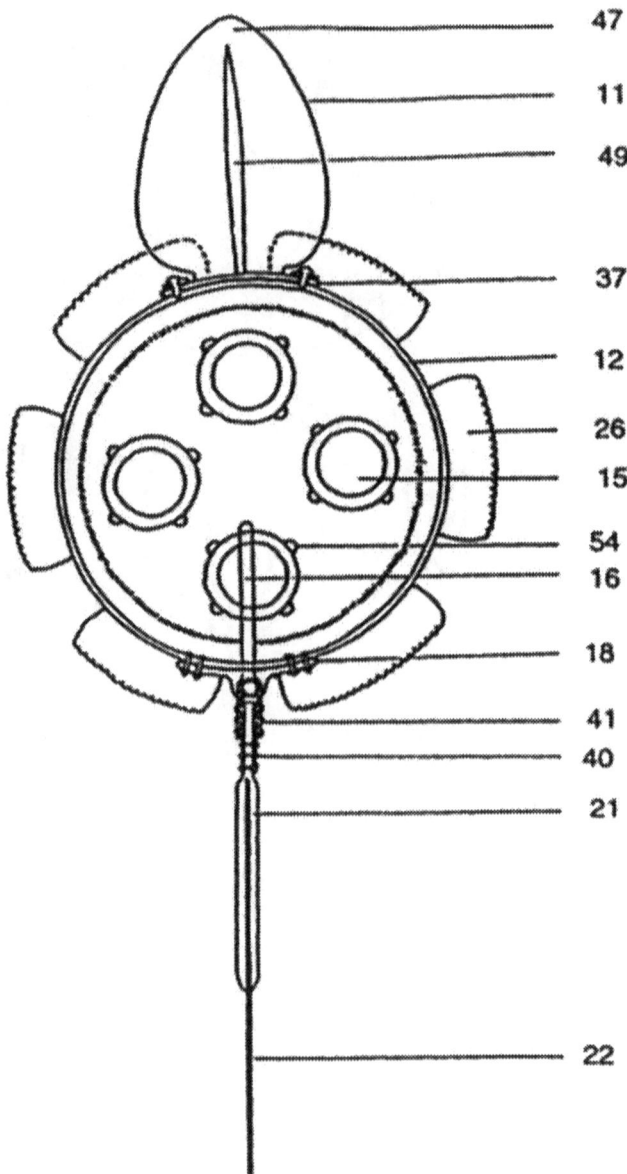

47

11

49

37

12

26

15

54
16

18

41
40

21

22

Fig. IV Showing the rotor-dinghy in top view

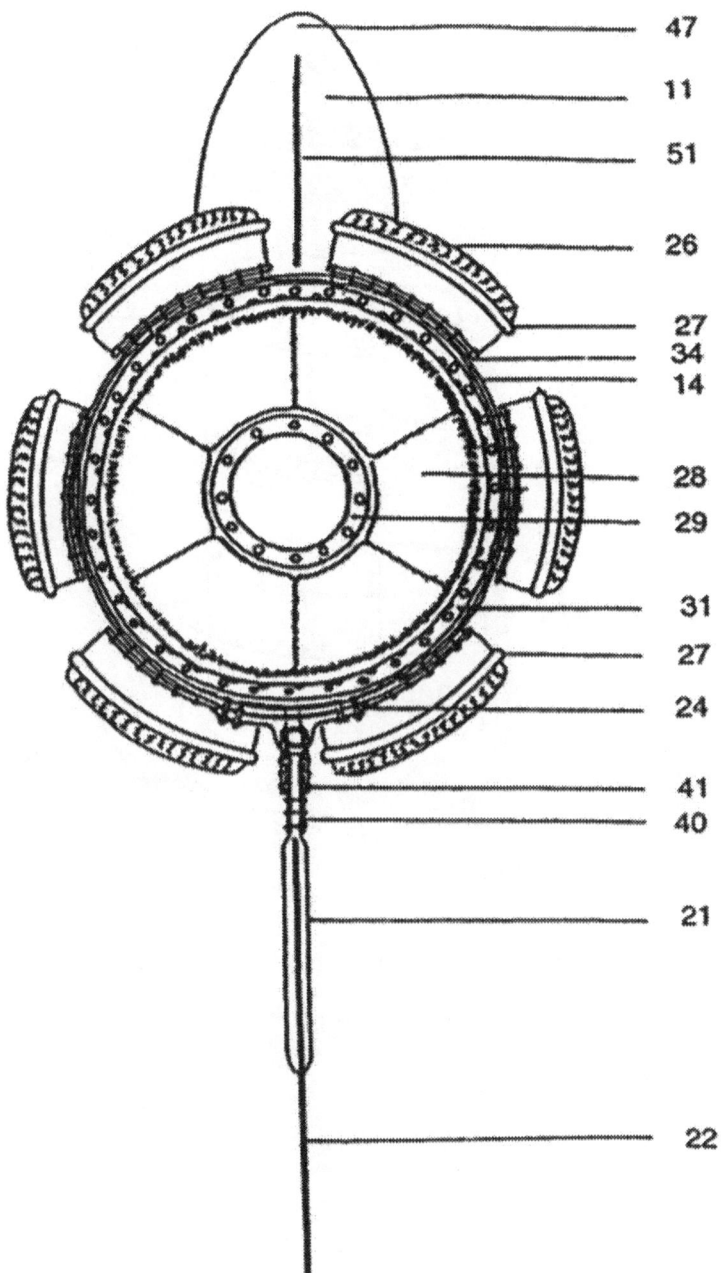

Fig. V Showing the rotor-dinghy in underside view.

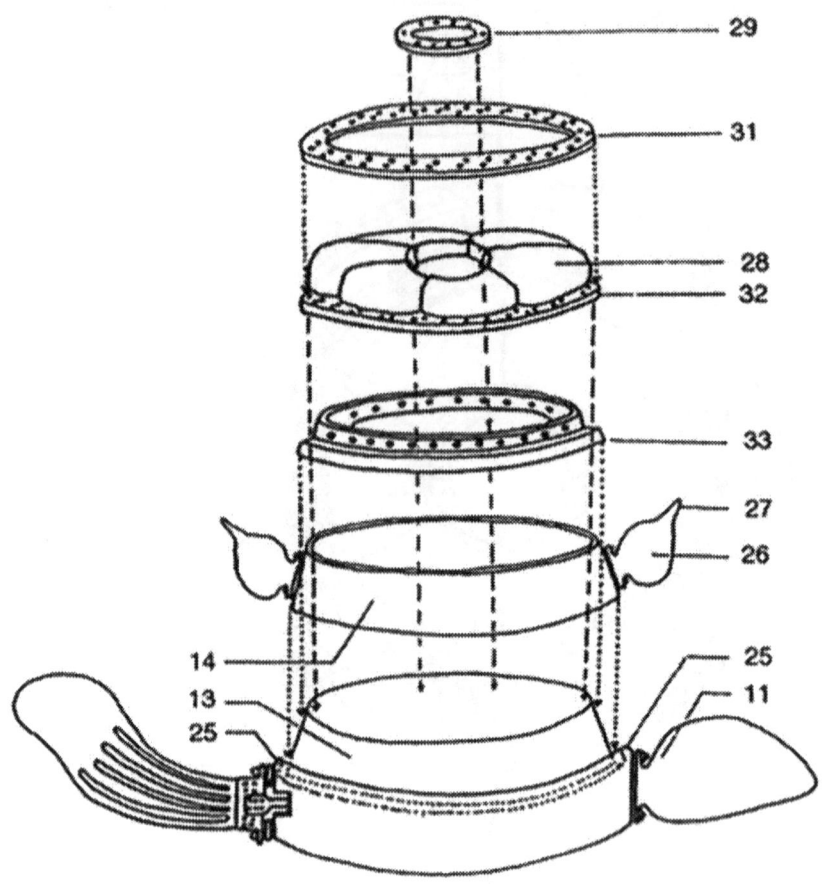

Fig. VI Showing consecutive stages in assembling the rotor-dinghy.

VIII. SHUTTLE BRIDGES TO OVERPASS OBSTACLES

Filing Date 10/4/99   Application No., 09/411,130

Foreign Filing License   Granted 10/27/99

A Concise Recountal:

The suggested shuttle bridge presented in this communication was triggered by the need of both Moroccan and Spanish governments to connect their countries to augment trading, traveling, and cultural exchanges. They were thinking of a 17-mile-long underwater tunnel that could run under Gibraltar Strait. The financial woes, the dangers of a collapsed underwater tunnel due to sea-bed tremors, and the difficulties of ventilating and getting rid of excessive heat produced by the traveling rigs made me think of another alternative. The envisaged, though in essence a suspension bridge, yet it deviated from the methodical. That is because the suggested shuttles carrying people and goods, will be running on its underside. This is accomplished by installing overhead wheels to the framework of the employed conveyances. The overhead wheels in turn are loosely hugged by kerfs acting as rails incorporated on the ventral side of the hoisted platform. This loose fitting of the wheels in kerfs help to suspend the shuttles or trains under the platform and at the same time allow them to move freely from one end of the bridge to the other.

PATENT APPLICATION

OF

WAGDY A. ASSAWAH

FOR

SHUTTLE BRIDGES TO OVERPASS OBSTACLES

Background Field Of Inventions:

Shuttles moving underneath a suspended platform were thought of as suitable means for connecting shores of relatively narrower straits, banks of rivers or over passing canyons. The contrivance though in essence is a suspension bridge, yet it deviates from the traditional. In the suggested shuttle bridge, a hung platform hoists in an unconventional way, the moving shuttles on its underside. See Fig. I. Meanwhile in traditional suspension bridges, the hung platform carries the roadways for the moving traffic on its upside.

One may look at this new envisagement as a cross between the traditional suspension bridge and the mountain cable gondolas.

Wagdy A. Assawah
Background Prior Art:

People have been building bridges for thousands of years. Fallen trees and rope bridges were used by our earliest ancestors. Nowadays, broadly categorizing, we have beam bridges for the use of trains, arch bridges, and suspension bridges. In arch bridges, support for their platforms comes from shored up subservient buttresses. On the other hand, in suspension bridges support for their platforms comes from bilateral steel cables strung from standing pillars.

Objectives And Advantages:

The concept of shuttle bridges with shuttles running on the underside of suspended platforms was triggered by the need of both the Moroccan government and the Spanish government to connect their countries. (Cited by The *Philadelphia Inquirer* of August 15/ 1999). Currently they are thinking of a 17-mile-long underwater tunnel that would run under the strait of Gibraltar. The financial woes of raising four billion dollars for the first phase is in itself a stumbling block, let alone the bother over mega-funds needed later on for its completion.

The idea of shuttle bridges connecting shores of straits, banks of rivers and traversing canyons is an appealing concept for several reasons. To mention few, engineers have mastered the building of colossus pillars to support platforms for procuring oil from the ocean floor. They also know how to protect them from the rages of nature, straying icebergs and the like.

Furthermore, the higher probabilities of dangers resulting in losses in human life, together with the exorbitant expenses for building an underwater tunnel out weighs those of building a shuttle bridge.

Also the ease with which to conduct effective control over incoming and out going shuttles at arrival and departure stations will be another appreciated advantage by the concerned authorities.

Last but not least, solo shuttles or shuttle trains, running in the open facilitates maintenance and makes rescue operations attainable ventures. Meanwhile, if conveyances together with their commuters get caught in a collapsed tunnel, they most probably, will be beyond salvation.

Description Of Drawing Figures:

The invention will be better understood when consideration is given to the following detail descriptions there of. Such descriptions make reference to the annexed drawings where-in.

Fig. I.   Showing a vertical section in a shuttle bridge with a suspended platform hung by strong steel cables. Also showing three steel slabs secured to its ventral side. The parallel edges of the slabs form U-shaped kerfs to accommodate the overhead wheels of shuttles traveling to and fro.

Fig. II.   Showing the shuttle bridges spanning over a strait connecting its two opposite shores. On water, the platform is hung by steel cables. On the two opposite shores, stout columns and subservient arches support it.

Fig. III. Showing in (A) an enlarged prospective of an omega railing, and in (B) a vertical section in this railing.

Fig. IV. Showing in (A) a shuttle in side view and in (B) a shuttle in top view.

Fig. V. Showing the shuttle bridge in top view. The part of the platform passing over the strait is hung by steel cables emanating form the apices of seven pairs of pillars and from gantries connecting them. Meanwhile, the parts of the platform built on land, form circular loops supported by stout columns.

Fig. VI. Showing a single lane uplifted by two opposite stout columns.

Fig. VII. Showing a double lane, as occurring immediately after a shunt, warranting a threesome of columns to uplift the bifurcating platform.

REFERENCE NUMERALS IN DRAWINGS.

11.   Platform.

12.   Pillars suspending the platform.

13.   Dorsal steel pane of platform.

14.   Ventral steel pane of platform.

15.   I-beams (8 in diagram) sandwiched and riveted to the steel panes.

16.   Omega railings (7 in diagram) secured to the dorsal steel pane.

17.   Steel cables reaching from apices of pillars and gantries to suspend the platform.

18.   Apex of a supporting pillar with the reaching steel cables hooking with the omega railings.

19.   A gantry connecting the facing pillars and with bowing steel cables hooking with the omega railings.

20.   Steel slabs (3 in diagram) secured to the ventral steel pane.

21.   U-shaped kerfs forged from the parallel edges of the steel slabs.

22.   Shuttle hanging by its overhead wheels and loosely hugged by the kerfs.

23.   Columns supporting the platform on land.

24.   Arches spanning between the columns for lifting the platform over land.

25.   Maintenance depot.

26.   Hump medially situated on the upside of the omega railing.

27.   Cleft beneath the hump.

28.   Apertures at equal intervals opening into the cleft.

29.   Limbs of the omega railings flanking its hump.

30.   Holes to pass the shanks for riveting the omega railings to dorsal steel panes.

31.   Free end of a steel cable terminating into a cleft.

32.   Cuff bigger than the aperture opening into the cleft to anchor the free end of a steel cable.

33.   Back door of shuttle for vehicles to get inside it.

34.   Front door of shuttle for vehicles to get out of it.

35.   Side doors for passengers to embark or disembark the shuttle.

36.   The overhead wheels of the shuttle.

37.   Jet engines for propelling the shuttle.

38.   Head lights of shuttle.

39.   The double loops of the platform reaching one shore of a strait.

40.   The double loops of the platform reaching the other shore.

41.   Inner smaller loop for a shuttle to continue on its returning journey.

42.   Outer wider loop for taking the shuttle to the maintenance depot.

43.   Shunts to shift shuttles from a smaller loop to a wider loop or vice versa.

44.   Arrival Station.

45.   Departure Station.

DESCRIPTION OF THE PREFERRED EMBODIMENT:

With reference now to the drawings and in particular to Fig. I, there-of the preferred embodiment of the new device embodying the principles and concepts of the present invention and generally designated by reference System 1 will be described.

Concepts Of System 1:

Specifically, it will be noted that the device related to connecting shores of straits, banks of rivers, or the steep palisades shouldering gorges and canyons. In the here-in presented envisagement, this is achieved by shuttles or shuttle trains hoisted and running on the underside of a suspended platform.

Concerning Fig. I:

The vertical sections in this figure is showing how the platform (11) is made of a dorsal steel pane (13), a ventral steel pane (14) with I-shaped beams (15) sandwiched between them. The three components, i.e., the two steel panes and the I-beams, are riveted together to form a coherent platform that can stand strong winds and also can resist sagging under the weights of the moving shuttles.

The platform is also helped to keep its posture by strong steel cables (17) emanating from the apices (18) of pillars (12) and bowing from the upper edges of gantries (19). The pillars in their turn are connected with gantries (19) to add to their strength and cohesion. The underside of the platform together with its hanging shuttles (22) should have enough height above the waters of straits or rivers to clear the loftiest of ocean going ships.

Riveted also to the underside of the platform (11), are three steel slabs (20) with their ends forming U-shaped kerfs (21). Each slab with its pair of kerfs form a lane for the running shuttles. One of the marginal lanes will be designated for shuttles running in one direction. The other marginal lane will be designated for shuttles running in the opposite direction. Meanwhile, the middle lane will act as an extra lane to be used for periodical maintenance or in case of emergencies.

However, builders of shuttle bridges may find it more plausible to have two emergency lanes instead of one. So, when need necessitates to maintain a marginal lane, shuttles of that lane have to be switched to an emergency lane. This shifting of a shuttle for a marginal lane to an emergency lane mandates the presence of shunts for the purpose. The shunts though cannot be represented in the submitted vertical section, are conveniently installed at the antithetical parts of the suspended platform spanning the passed over strait and before the marginal lanes bifurcate forming the circular loops (39 and 40) built on the two opposite shores. (See Fig. V.)

Complementary Components For System 1

System 1 presented here-in, is complemented by various other figures, namely II, III, IV, V, VI, and VII to disclose other features and to explain their uses and performances.

Concerning Fig. II:

The side view in the figure is showing a shuttle bridge spanning over a strait and connecting its opposite shores. Over the strait, the platform (11) is hung by canopies of strong steel cables bowing from gantries (19) and apices of the supporting pillars (12). In the diagram, there are seven pairs of pillars that establish their foundations in the sea bed, then rise above the sea and subsequently suspend the shuttle bridge.

On the other hand, the parts of the platform reaching the opposite shores are supported by stout columns (23) and arches (24). The latter, i.e., the arches help in keeping the platform supported above the open spaces between the columns. The opposite extremities of the platform reaching land proceed into maintenance depots (25) where if needed, periodical oiling, fueling, cleaning…etc. are carried out for the operating conveyances.

Wagdy A. Assawah

Concerning Fig. III:

The drawing in part A of the figure is showing a prospective view of an omega railing (16). While the drawing in part B of the figure is showing a vertical section in that railing. Both drawings show the detailed structure of this component and show how the omega railings help in suspending and securing the platform in place. The vertical section of the shuttle bridge given in Fig. I is showing seven of the omega railings (16) extending the whole length of the suspended platform (11). There could be more needed, if the builders choose to have four lanes instead of three.

The omega railings are firmly riveted to the dorsal steel pane (13) of the platform (11) by means of passing shanks in the holes (30) made in the oppositely flanking limbs (29). The omega railing takes its name form the dome-shaped hump (26) located on its upside and from its flanking limbs (29), reminiscent of the Greek letter omega. Each hump harbors underneath it a cleft (27) that opens to the outside by lined-up and equally distanced narrow apertures (28).

The vertical section presented in part B of the figure is showing a free end of a steel cable (31) terminating into a cleft

(27). In there the end is steadfastly held and detained by a cuff (32) bigger in size than the aperture (28) made in the hump (26). These detentions, when brought about by all the cuffed ends of the steel cables, grant a firmly suspended platform that can stand the forces of wind gales and the weights of moving shuttles.

Concerning Fig. IV:

Though Fig. I reveals the inside of shuttles and how they can accommodate passengers and vehicles, yet it was necessary to display them from other angles of perception. Hence, to satisfy this need, drawings A and B of Fig. IV were incorporated to show respectively a side view and a top view of an operating shuttle.

Atypically, these shuttles are provided with overhead wheels (36) located on both sides of the upper part of their fuselage. The wheels are to be accommodated in the kerfs (21) of the overhead slabs (20) lining and secured to the ventral pane (14) of the platform.

Each shuttle has a back door (33) for cars and other vehicles to get inside it and a front door (34) for them to get out of it. For

Wagdy A. Assawah

passengers to embark or to disembark at arrival stations (44) or to get off at departure stations (45), there are side doors (35) for the purpose.

Needless to say the shuttles can be driven either by electricity or by jet engines (37). And like any means of transportation it is outfitted with headlights (38) and taillights.

Concerning Fig. V:

The presented here-in figure is showing a bird's eye view of the shuttle bridge. It displays the whole gamut of the three parts comprising that bridge; fitting, working together and forming a continuum. These three parts are; the suspended platform (11) forming the middle part spanning over the traversed strait, and the two end parts reaching the opposite shores forming double loops (39 and 40). These double loops emanate from the antipodal ends of the suspended platform (11).

The platform suspended over the strait is held by a canopy of steel cables (17) coming down from both the apices of pillars

(18) and their connecting gantries (19). On the other hand, the parts built on land are suspended by stout columns (23) and arches (24) spanning the spaces between the stout columns. Both double loops (39 and 40) form on each corresponding shore, an inner smaller loop (41) and an outer wider loop (42). The inner smaller loop (41) accommodates a single lane to allow the shuttle to continue on its returning journey. The outer wider loop (42) also accommodates a single lane to take the shuttle to the maintenance depot (25). The splitting into two lanes is brought about by incorporating shunts (43) to allow the shuttles to jump lanes and to go form the smaller loop (41) to the wider loop (42) or vice versa.

It may be worthwhile to mention that these shunts switching the shuttles from the smaller loops to the wider loops and vice versa are other than those located at the antithetical endings of the suspended platform helping the shuttles to shift form a marginal lane to an inner emergency lane.

Wagdy A. Assawah

Concerning Figs. VI And VII:

The two diagrams represented by these figures show how the platforms and their hoisted shuttles are depicted on land. Platforms on shores are supported by stout columns (23) and arches (24).

In Fig. VI. The platform is uplifted by a pair of columns (23) with a steel slab (20) lining and secured to the ventral pane (14) forming a single lane. With their overhead wheels (36) accommodated in the kerfs (21) of the steel slab (20), a shuttle will either be on its returning journey or on its way to the maintenance depot.

Meanwhile, Fig. VII. is showing two adjacent lanes uplifted by a threesome of columns. This is a result of a nearby shunt (43) splitting the original lane in two. One lane will comprise the outer wider loop (42) proceeding to the maintenance depot (25), and the other will comprise the inner smaller loop (41) proceeding to the suspended platform and eventually to the other side of the traversed obstacle.

OTHER INTRINSIC POTENTIALITIES OF THE
ENVISAGEMENT

While the above description contains many specifications, these should not be construed as limitations on scope of the contrivance.

The ambitious jet-propelled double-decker shuttles shown in Fig. I could be replaced by shuttle trains towing several cars. Yet to conform with the job, the locomotive and the towed cars has to have overhead wheels installed on both sides of the upper parts of their framework to fit the lanes hoisting them.

Adding tiers of gantries to connect facing pillars establishes a multiplicity of steel cables bowing from the added gantries. This multiplicity assists those cables emanating from the upper most gantry to reach out for the farthest halfway locations befalling in between two consecutive pairs of pillars. This potentially allows building longer shuttle bridges with fewer pillars, as may be needed for passing over gorges or narrower canyons.

Wagdy A. Assawah

Variations could also encompass the width of the platform to cater for several lanes for both emergency purposes and for the journeying means of transportation. These could either be, solo shuttles, shuttle trains, or both, with each conveyance utilizing its own concordant lane.

PAGES ENCOMPASSING THE DRAWINGS

WITH THE FIRST PRESENTING

SYSTEM 1

AS THE PREFERRED EMBODIMENT OF THE

"SHUTTLE BRIDGES"

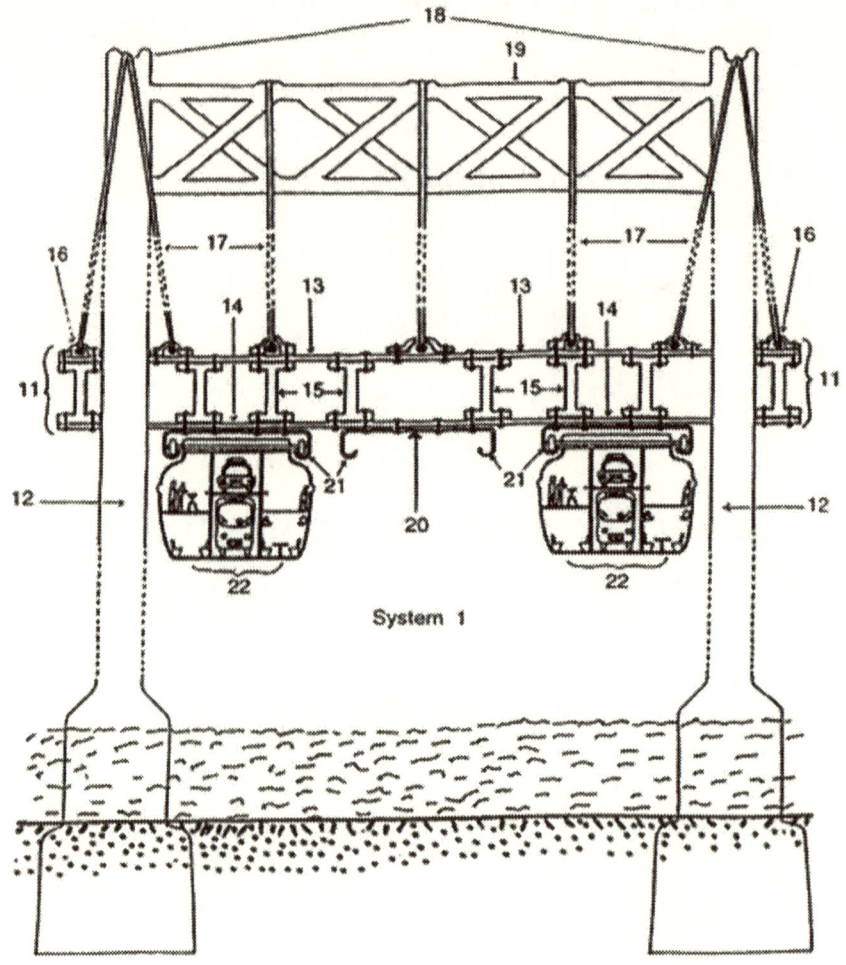

FIGURE I Showing a vertical section in a shuttle bridge with a suspended platform hung by strong steel cables. Also showing are three steel slabs secured to its ventral side. The opposite edges of the slabs form U-shaped kerfs to accommodate the overhead wheels of shuttles traveling to and fro.

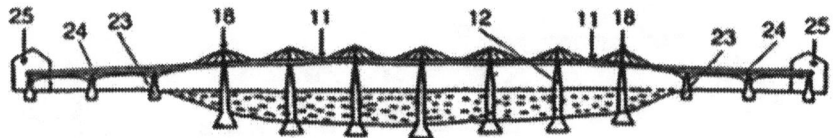

Figure II Showing the shuttle bridge spanning over a strait connecting its two opposite shores. On water, the platform is hung by steel cables. On the two opposite shores, it is supported by stout columns and arches to keep the platform above the openings between the columns.

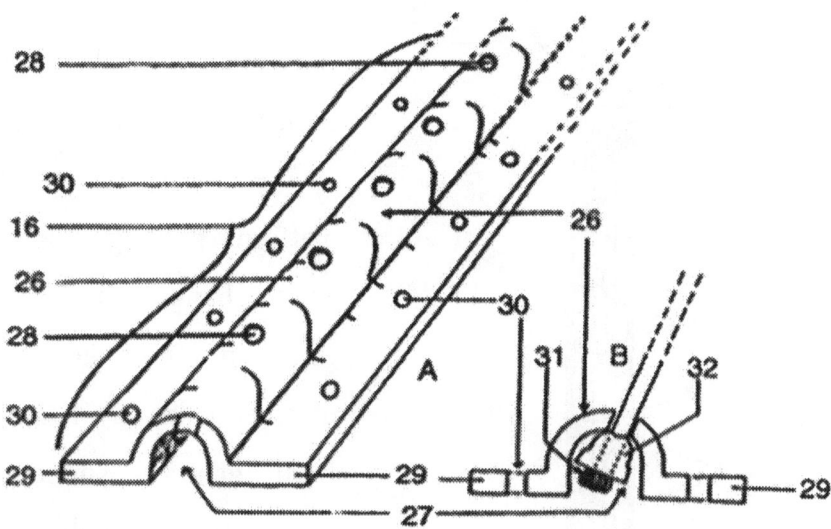

Figure III Showing in (A) is an enlarged prospective of an omega-shaped railing. Shown in (B), a vertical section in this railing.

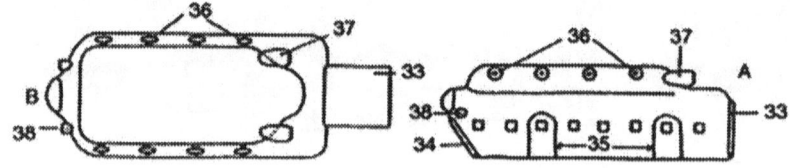

Figure IV Showing in (A), a shuttle in side view. Showing in (B), a shuttle in top view.

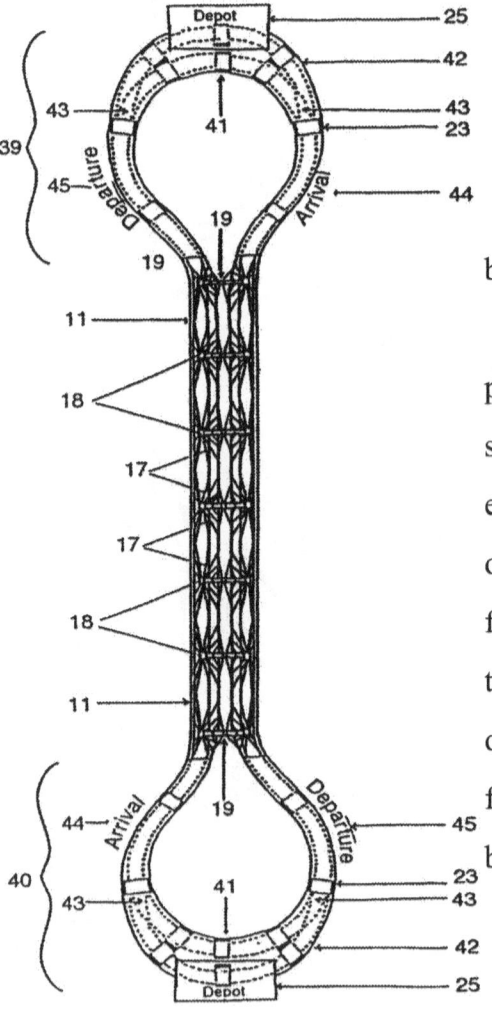

Figure V

Showing the shuttle bridge in top view.

The part of the platform passing over the strait is hung by steel cables emanating from the apices of seven pairs of pillars and from gantries connecting them. Meanwhile the parts of the platform built on land form circular loops supported by stout columns.

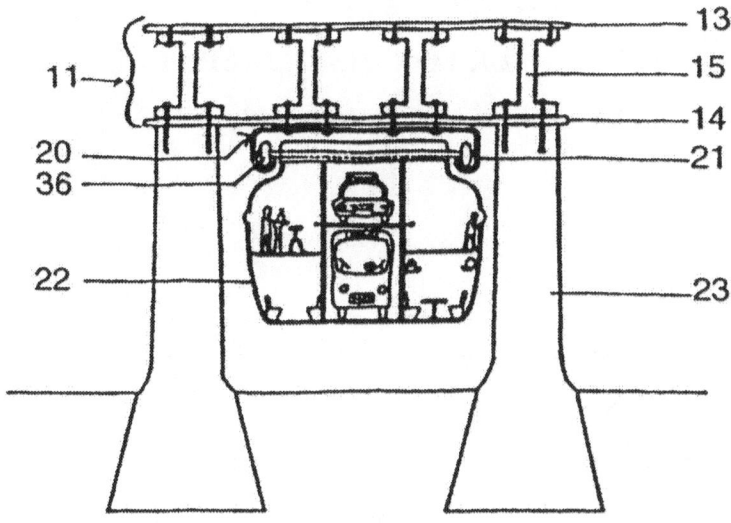

Figure VI

Showing a single lane uplifted by two opposite stout columns.

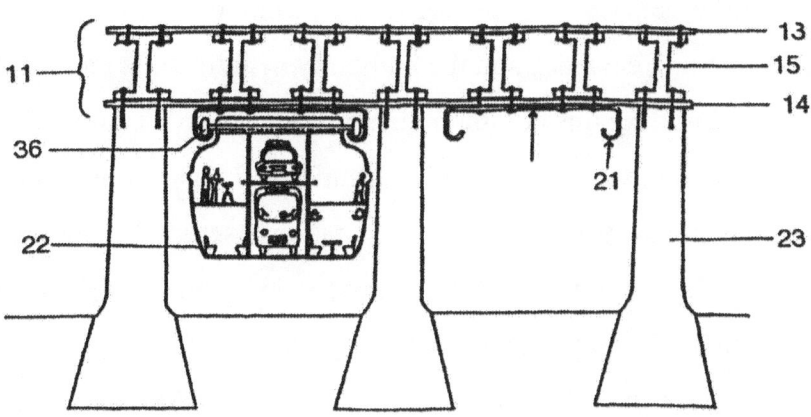

Figure VII

Showing a double lane as taking place immediately after a shunt warranting a threesome of columns to uplift the bifurcating platform.

IX. THE MANUMITTER
A SURE SAVER FOR DISLIMBED PLANES
AN ULTIMATE MEGA-MACHINE

By

Wagdy A. Assawah*

Prologue:

In the past few months of 2005, the media showed millions of their viewers, several flitting planes that were unable to make safe landings. That was because their pilots were unable to lower their jammed landing wheels. Now to order the pilots to fly around for several hours in order to get rid of the plane's fuel, and to make a belly landing in a nearby field is no plausible solution. Also to keep the scared commuters and their relatives and friends waiting on pins and needles is by the same token, inhumane. It was these life threatening episodes and the malignity in action that enticed the author, submitting this communiqué, to think of a plausible modus-vivendi to bring such disabled planes and their human commuters to safety. Doing so, the author believes that he is setting a stone in the building of the world. And that his contrived "Manumitter" is none other than an authentic safety delivering modern marvel.

Now without further ado, the following two drawings are here to show how the components of this newly introduced "Manumitter" are integrated. And to show how they work effectively to insure safe deliveries for both the trusting passengers and their overworked planes.

* *Sworn and subscribed before me This 13th day of December 2005 Martin La Venegas, Notary Public Falls Twp. Bucks County. My Commission Expires Sept 9, 2006.*

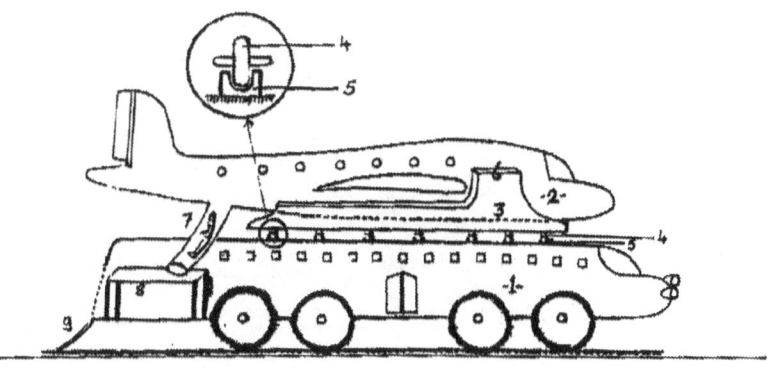

Fig. I: Showing a side view of "The Manumitter (1) and the hugged dislimbed plane (2) resting on the flattened back of the Manumitter

To keep the plane into a clinch, it is hugged by two longitudinally flanking wing-like structures forming a bivalvate cradle (3). The padded-on the inside cradle halves, can move sideways, to either open wide to accommodate the homing-in plane (2), or to close up to keep the plane in place. Their getting apart from each other, or their getting closer is due to wheels (4) fitted to their underside running in grooved railings (5). The railings are numerous, equidistantly fixed on the flattened top of the Manumitter. Also, to prevent the plane from moving forward, each cradle half elongates upward to form an impediment (6) to stop the plane from going further.

Once the plane is in place, the passengers can use the chute (7) to safe land on a trampoline (8) inside the Manumitter. From there they can go to slide on a glissade (9) fitted to the back door of the Manumitter to enjoy the euphoric sensation of standing once more on solid ground.

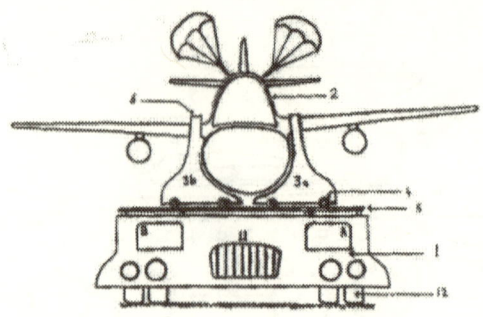

Fig. II: A frontal view of "The Manumitter"

The Manumitter (1) is a sturdy, stubby mega-machine. It has a powerful engine warranting a huge radiator (11). It is also equipped with elephantine wheels (12) that can stand its own weight and the weight of the salvaged plane (2). The Manumitter need two skilled operators. One in window (A) for driving this colossus, while the other in window (B) is for working the cradle halves dynamism.

In the presented here-in drawing, the Manumitter is carrying the salvaged plane between the cradles halves (3) shown here by a pair of structures depicted in the vertical sections (3a) and (3b). Helped with their wheels (4), the cradle halves can go opposite ways to accommodate a homing-in dislimbed plane. With the plane settling between the cradle halves, it is then hugged to keep it in place. Also with the aspect of a protruding impediment (6), the plane will be prevented from going further. This can like wise be helped by releasing the drag-chutes.

Together with the manifested earlier, the Manumitter is capable of attaining very high speeds to allow it to synchronize with the pre-stalling speed of a homing-in dislimbed plane. Safely in the cradle halves, the plane can then free its commuters, and if need be, the plane can be taken for inspection and repair.

PART IV

APPENDICES

If I have seen further, it is by standing on the shoulders of giants.

Isaac Newton

APPENDIX I.
ABOUT THE AUTHOR

Right from the start I refrained from writing about myself. I believed that the message is more important than the messenger. But now I feel indebted to respond to the question reverberating in the subconscious of the readers, and to show that the podium is not a strange place for me.

The Molding Of my Personality:

I was born on December 31, 1924 at Port-Said a small town at the northern tip of the Suez-Canal. The promoter of this waterway was the French architect Ferdinand de Lessepse. It was opened by Empress Eugenie in November 16, 1869, and Port-Said was built by the French Company managing this important waterway connecting the East with the West.

At Port-Said High School, where I was educated in the early 1940's, I took part in writing articles in its magazine. I also supervised the activities of the "Scientific Society". As a group we made different electric circuits to ring bells, to switch on and off bulbs, horseshoe magnets from bent nails, crystals of metallic salts, osmotic flower gardens, coat a base metal with a noble metal, enjoy color changes in different indicators…etc. We were also the happiest when our handmade balsa wood airplanes took to the air. Among friends my activities earned me the moniker, "The inventor".

Also, our classrooms were generously decorated with portraits of scientists, philosophers, inventors, writers and other universally known erudites and patriots. I looked up to them hoping one day, to walk in their boots. This zealotry was recognized in me by my father who instilled in me that the best men are those who take upon themselves to make of this world of ours a better place for his fellow men.

My Education And An Allied Contingency:

After finishing my high schooling at Port-Said in 1944, and with my ambitious attitude, I went to King Fouad University at Giza to get a higher degree from the Faculty of Science.

In 1946, I took part in the riots against the paid subversive elements supplying the battle against the Israeli Forces, with tainted ammunition that resulted in a humiliating defeat for the Egyptian Army. On this infamous day the police forces under the command of their British Officers followed the rioters to Khedive Abbas Swing Bridge, the way to center city Cairo. At Giza's end of the bridge, the trailing police forces were ordered to fire on us. Meanwhile, at Cairo's end of the bridge, and unknowingly to the rioters, the bridge was swung open. Thus we were trapped between a death by bullets or a death in the eddies of the Nile. Consequently some were deadly wounded while others in the chaos were either stampede or pushed over the open edge of the bridge to die drowning. In a hand* combat I got a fractured skull and a cut on the nose from the buttock of a rifle. After I was treated from my wounds I was incarcerated for two weeks until the demission of the sitting government. With the coming of a new administration, all the jailed students were exonerated and freed.

In the summer of 1949, I was awarded an honorary B.A. degree in chemistry and botany. In the graduation ceremony the distinguished few, and I was one of them, were privileged to be congratulated by King Farouk and to shake hands with his majesty. Looking for a career, I accepted a teaching position at the New King Farouk University at Alexandria. There I spent two years as

*This was my Tiananeman Square moment

an associate lecturer teaching the practical lessons in biological laboratories and doing research to get an M.S.c. For lack of appropriate equipment at the new university, my supervisors at the Department of Botany nominated me for a sabbatical leave to continue my studies abroad. Though my father wasn't Jacob nor was I Joseph, he dreamt that I was going to visit with King George VI. My father's dream, or for that matter his well wishings for me, was a good omen. Surprisingly the United Kingdom was chosen to be my destination and in October 1951, together with my newly wedded wife, we boarded the P.&O. Chusan from Port-Said to arrive at Tilbury on the River Thames.

At Cambridge University I studied for two years as an undergraduate and in 1953 I got my B.A. Hons in Botany. From Cambridge I went to Nottingham University to do my doctorate in Plant Pathology under the supervision of Prof. Dr. Charles G.C. Chesters. My submitted thesis awarded me a Ph.D. from Nottingham University in 1956.

During my sabbatical leave in England, which extended from 1951 to 1956, General Mohammed Naguib in 1952 seized power and King Farouk was forced to abdicate. In 1954 Colonel Nasser replaced General Naquib and became the Head of State. Consequently King Fouad University at Cairo, was renamed Cairo University and King Farouk University at Alexandria became Alexandria University.

Pursued Careers:

Returning to Egypt, I was appointed at the Faculty of Agriculture at Alexandria University as a lecturer from 1956 to 1962, then as an Associate Professor from 1962 to 1967. At this college together with supervising postgraduates doing research to get M.S.c. or Ph.D. degrees, I also lectured in Virology, Bacteriology, Mycology, Plant Pathology, Insect transmission of phytopathogens and in Industrial Microbiology. I also published more than 40 scientific papers on phytopathologial problems in several reputable journals domestically and nationally. My published papers earned me in 1967 a professorship at the Faculty of Science at Tanta affiliated to Alexandria University and to be the chairman of the Botany Department from 1967 to 1976.

In 1976, becoming a tenured faculty member, the chairmen of the different departments at the Faculty of Science, i.e., Physics, Chemistry, Mathematics, Geology and Zoology elected me to be the Head of this Scientific Institution from 1976 to 1978.

Gained Experiences:

• Holding educational key positions in universities subjected me to a multitude of duties and experiences. During my extended career and under my supervision, 23 post-graduates were awarded M.S. degrees and 11 were awarded Ph.D. degrees. A number of the completed theses were made in cooperation with European Universities under what was called, "Research Channels". A post-graduate at the department will do part of his work in Egypt and the other part abroad. One of the promoters of this endeavor to whom I owe much gratitude was the late Professor Dr. W. W. Fletchers, Head of the Department of Biology at Strathclyde University, Glasgow.

- To keep the efficiency of the labs and green houses, in spite of the shrunken finances over the Egyptian-Israeli conflict, I did my best to provide them with the needed equipments. Thus apparatuses, glassware and chemicals were provided for the different labs doing research in Microbiology, Plant Physiology, Cytogenetics, Plant Taxonomy and Ecology.

- I was a member in several scientific societies, The British Mycological Society, The Egyptian Botanical Society, The Italian Phytopathlogia Mediterranea, The Applied Microbiological Society and I was the treasurer of the Egyptian Phytopathological Society.

- I was the chairman of "The Committee of Botanical Sciences" for examining submitted papers by lecturers applying for Associate Professorship status. And I was a member of the corresponding committee for examining submitted papers by Associate Professors applying for Professorship status.

• I refereed a great number of scientific papers for Journals in the fields of Microbiology, Mycology and Plant Pathology and examined uncountable numbers of M.S.c. theses and Ph.D. theses to award their successful candidates the appropriated degrees.

• To enrich our Arabic Library, I co-authored a book on General Botany and wrote a second on "Diseases of fruit crop trees and their control" and a third on "Diseases of flowering and ornamental plants".

• As an emeritus professor participator, I taught at the Department of Botany at Mosul University-Iraq, for the academic year of 1981-1982. Also, I taught and supervised research at the Institute of Biology of Oran University, Algeria for three years from 1982 to 1985.

• Going on pension in 1985, I was awarded a Medal of Distinction for my humble scientific services and for establishing several biological labs. The medal was thus the crowning of all the work and toil for years that has gone before.

Coming To America:

The disappointment and the resolution.

Becoming a pensioner in 1985, my wife and I decided to join our American daughter, Rhoda, who graduated in 1983 as a pharmacist from Temple University at Philadelphia and to join our American son, Nader, who graduated in 1984 as an artist from the Art Institute of Philadelphia. In 1993 my wife and I were admitted to be citizens of the United States of America.

In this great country of ours, I thought that my British degrees, my experiences and my Medal of Distinction from Egypt would guarantee me a decent job. So I started applying for jobs in research facilities. Yet I was always confronted of being over qualified. For keeping active I worked as a substitute teacher in nearby schools. Meanwhile my wife took part in the preschool programs organized by our local YMCA.

However, the continued refusal did not dampen my spirit. This gave me ample time to think about certain of our contemporary problems and to design gadgetries to lessen their harms. With a number of them rounded up, and together with certain opinions from

457

encountered experiences, I decided to publish the whole gamut in a book form hoping it will do good by my new country in particular and by humanity in general.

Hence if there are imperfections - and certainly there will be - blame it on my undue confidence that unregrettably led my fervor. Simply I decided to be the phoenix rising from the ashes to claim a worthy past and to continue a legacy of reproductivity. And to confirm to my beloved America, that I did not come here to retire but to be humbly useful.

Happy to conclude my work, the questions I keep asking myself, will I live long to know the outcome of my contributions? Will I go down in the records as a visionary? Or will I go down as another Don Quichotte who took upon himself to fight the windmills to rid the world of their evil? Whether this or that, my contributions for what worth they may be, were done in good faith and with a noble intent. And though I may fall short from reaching the Holy Grail, I must admit that I took pleasure in the journey.

APPENDIX II.
PROPOSED SUGGESTIONS

Perhaps I may be one of few privileged to look into the field of inventiveness in its totality, i.e., its materialistic gadgetries, its spiritualistic convictions, and its intellectual ruminations. Consequently, the responsibility of knowing better, obligated me to come up, to the best of my ability, with an unbiased opinion to uplift our inventiveness per se to a higher level of betterment. Therefore, I propose that besides investigating the submitted inventions, the P.T.O. should single out those promising inventions of national bearing and those of humanitarian importance—usually overlooked by the industry—to be studied thoroughly. If serious scrutinization reveals that they are endowed with undoubted benefits and capable of making a difference, then such inventions shouldn't be left to wither on the vine. Logic corroborates that we shouldn't let good deeds go unexplored, and to make good use of them. In fact, I see nothing more abhorrent than ignoring a fruitful invention on the assumption that its responsibility rests within its originator. This is especially so if the inventor hasn't got the means to manufacture its prototype, let alone producing it en masse. And I see nothing more exalting and rewarding in the job of examining and judging inventions than to single out a sure winner.

Our U.S.A. is a great nation, if not the greatest on earth, thus it is destined to lead by example. With God's given riches, this nation of ours that can afford billions to spend on interplanetary projects and to make and improve assortments of war machines, can also afford a small budget to take care of such promising peace promoting inventions for a change. So to execute such inventions, which we may need to build, to test, and even to improve, I suggest establishing a new, financially independent **"Governmental Commission for salvaging the worthy inventions."** This long awaited for organization, due to its undoubted importance and due to our unwritten moral commitment of giving a helping hand to underdeveloped countries, shouldn't delay any longer. My advice with the dawning of the twenty-first century is to open wide the doors of this new agency because many things that need to be invented hasen't been invented yet.

If Ever Inventions Can talk, What Will They Say!

Intuitively listening in, they will be saying:

- We are on the rise and we will be flooding the concerned institutions to the point of paralyzing them.

- There will be a majority of trivial inventions submitted by the greedy targeting a fast buck.

- There will be few genuine inventions that need insighted examiners to prioritize them, because of their value as life-changing endeavors and because they intend to benefit mankind without targeting the fast buck.

What can we do then to protect the good inventions from being lost in the deluge of the bad ones?. Certainly we cannot afford to lose them. We also need them to bolster our economy and to benefit mankind.

And if ever inventors come together, what will they ask for!

Certainly, they will ask for eliminating the middle man. For a reminder, inventors will neither forget the exploitation of "The Corporation Services" which unfortunately, can use their potency to patent the unworthy. Nor will they forgive "The Brokers" for spoiling the climate between them and the industry on one hand and for causing losses in both jobs and in products on the other hand, subsequently hurting our economy.

What can we do then to lessen the harms brought about by those exploiting organizations?. In fact, asking for putting them out of commission will be a tall order, yet a plausible one. In a world teaming with inventive minds and over flooded by invention manuscripts, we have to roll with the punches and to develop new ways to sort out the good that will benefit the people and to exclude the bad that will waste our time. Believing in taking a straight line to where our target is, and believing that this is the best policy, I allowed myself to suggest the following:

- Setting up an intermediary workplace, under the umbrella of the P.T.O. to publish only, for a fee, the plausible inventions as is, yet in perfect English in monthly issues. These issues are meant to reach the interested and the Industrial Enterprises. In so doing, we will be exposing the worthy inventions and putting them under the critical eyes of the knowledgeable in every art. Sifting through the issues, the industry can then choose what will be lucrative to the economy, and leave out what is not.

- The chosen invention is then sent by the concerned firm and with the blessing of its author, to the P.T.O. to be patented for the inventor. Happy to know that his invention has jumped the gauntlet, he will give his authorization to this particular firm to introduce it to the market. Needless to say, most earnest inventors will be appreciative upon getting the approbation of the industry and will be uncritical about the royalties. And I know that I am one of those earnest inventors. The royalties, for an obvious reason, were usually exaggerated by the brokers, making of it a stumbling block for the inventors and consequently losses in the National Revenue.

The earlier mentioned suggestions, stress several points of view with the aim of clearing the air between all the concerned in the field of inventiveness.

And the followings are for highlighting such proposals.

- Creating an "Intermediary Workplace" to select the plausible inventions and to expose them to the industry is of great importance from a systematic point of proper procedures.

- Sorting the good inventions and protecting them from getting lost amid the unworthy ones will be like preventing the bad coin from out-crowding the good coin.

- Eliminating the middle man and letting the industry to choose what is worthy for promoting, will be beneficial to the National Economy and will provide needed jobs to the workforce.

- Establishing trustworthy relations between the industry and the inventor will be an exemplary attitude worth encouraging for goodness of the profession and for a healthy International Trade.

To conclude and to show that the suggested simplicity is for the goodness of all concerned, here-in via my published book I am exposing what I can offer, to those who can make it, for those who can benefit from it. And to show that doing business on the personal level, without the intervention of the middle man, if not a win win situation, is at least a route worth while trying.

Hence, I welcome getting in touch with the industries wishing to manufacture any invention of mine. I am also open to other enterprises wishing to make use of my ideas, namely producing the minimal caloric foods, filming the rendezvous of the two antithetical time travelers to nihility, or making educative episodes based on leafing **The Book of Genesis,** starting with either **The Big Bang** or **The Big Crunch**.

Inventions As Weapons Against A Sagging Economy

Inventions have also benevolent sides. Warranted by the crash of 2001 in the financial market, it is worthwhile to point out that establishing new industries to manufacture new gadgetries can be used as effective means in combating debilitating recessions. However, if they cannot bring back a bygone financial boom on their own, they will at least put brakes on an economical doom. Merely by offering new jobs to an idle workforce laid off by bankrupted companies on one hand, and by providing a bonanza of useful products on the other hand, the new gadgetries will definitely be a contributing factor in reinstating a lost prosperity.

Corroborating the above presumptions, an ironclad advice was solicited from a called upon guru. You may, if it pleases you, call him, **"The Recession Buster"**. (Abbreviated **Mr. R.B.)** He raised his green thumb upward and said, **"Haven't you learned from past recessions! It is jobs and more jobs, stupid."** Murmuring up my sleeve, "Certainly the stupid are those who ignore your valuable advice, Mr. R.B."

Valuing our inventors and their useful inventions will boost our national egoism. Also, showing our interest in inventions that solve problematic hardships for the technologically deprived will be a source of our bride. In my belief, it is not satisfactory to exhaust our lungs singing "America, America, God shed his grace on thee", unless we show grace to the deprived and together with it we may be appropriately called, **"The Comely Americans".**

Now the question is "Can we administer the earlier mentioned suggestions?" And since progress is impossible without change, and those who cannot change their minds cannot change anything, the answer is an Emphatic "Yes".

APPENDIX III.

AD INTERIM

In Recognition Of A Shared Congeniality*

This is a spotter of headworks adducing life-changing inventions and other worthy suggestions allotted in this compendium. It is intended for the eclectical reader who flips the pages à posteriori to find out for themselves, what the book at hand is all about.

Judging a book by its cover, or for that matter, an author by his name, will continue to be a dubiety. The inaugurated herein "Ad Interim" is an eventuated resolution to unravel this mystification. However, since a novel is different from a scientific endeavor, thence to include or to exclude an "A.I.," should be bequeathed to the author's gumption.

CONCERNING THE MENTIONED INVENTIONS

Inventions For Desalting Sea-Water: (Page 209 & Page 224)

Two gadgetries were designed to secure fresh water from the seas to benefit man, beast, and cultivated land. Their assembled parts make them the closest devices to **"The Perpetual Motion Machine"** dreamt of by inventors throughout the ages. And make of them the fancied apparatuses favored by ecologists, since they are environmentally friendly.

The yield of the prototype may help in growing crops, if not in the field, at least in hydroponics. Improvements to increase the yield, is an eventuality for turning out second, third, and future generations till we reach the "Gregorian Threshold".

Understanding their importance to nations suffering from shortages in fresh water, manuscripts were sent to Egypt and Israel. For Egypt, to enable them to reclaim the lost fertility of the strip bordering the Mediterranean Sea. For Israel, to head off a possible war between neighbors over the insufficient income of the "Jordan

River". Expectedly, if one country or the other, utilizes my apparatus or a derivative of it, I hope to be credited for the artistry and the concepts incorporated in its contrive.

The Manually Driven Water Propulsion Device: (Page 243)

It was invented to introduce a new water sport. This water propulsion device "W.P.D.", enables its user to move around on water while lying on his back.

With reference to inventions dubbed **"Exo-Skeletor, Solo-Trek, Human Performance Augmentation Devices",** the W.P.D. is one of the earliest mechanisms incorporating these concepts. It preceded Mr. Michael Moshier' "Flying Vehicle" and Mr. Dean Kamen' "IT" which operates on dry land and was televised for the first time on December 3, 2001.

While the flying vehicle was designed to take off and to land like a helicopter to help soldiers to get around obstacles, the W.P.D. can help a navy seal to get around in harbors in a stealthy manner to surprise the enemy. The inventors wrote to the U.S.A. Navy about its capabilities. Unfortunately, our communiqué neither caught their attention nor did it arouse their curiosity. Now with more exposure, hopefully the industry can introduce it to the dare-devils of water sports.

Protected Heliports To Safeguard Commuters: (Page 269)

These heliports were invented to safeguard our valuable leaders, our moguls of leading industries, and our prodigious entertainers, from peeping Toms, insistent paparazzi, or worst from menial snipers. Thus, our valued elite can leave their offices or homes and come back to them in complete privacy, fearing neither a shot from a camera nor a shot from a treacherous assassin.

Evulsion Mechanisms To Sever Skin Flaps: (Page 304 & Page 343)

First came the evolusion clamp, they were invented to sever the unwanted skin flaps. Being non-surgical procedures, the treated patients can stay in the comfort of their homes till the ousting of the fold is actuated.

However, unsatisfied with the evulsion clamps as suitable means for severing the bulky belly flaps, I strived to come up with a better procedure closer to the edge to do away with such skirting flaps. In this second generation, the evisceration is done by "Snare Tackles". For their distinctive ability to rid us from such cumbersome flaps, there is no doubt in the author's mind, that they offer a short cut for winning "The Battle of the Flap".

The clamps and the snare tackles will undoubtedly change our way of life. They will introduce a new medical profession that will be practiced by authorized specialists. Comparable to chiropractors they can be called, "The Evulsopractors".

A Rotor-Dinghy For Shooting Rapids: (Page 388)

The suggested dinghies are provided with revolving wheels hugging their basal parts. In their turn, the wheels are shored up with inflated bolsters. This design was borrowed from mother nature, who knows best. Like the mantels of jelly-fish, the inflated bolsters shoring the revolving wheels, spread out to cover a wider circumference to ease a dangerous descent, and to autofold to affect a quick run-off, allowing for a gentler ascent.

Shuttle Bridges To Overpass Obstacles: (Page 421)

The need to build a 17 mile long underwater tunnel under "Gibraltar Strait" triggered the idea of making a, Shuttle bridge as a plausible alternative. This latter will not need the continued and costly expenses of getting rid of the excessive heat produced by the traveling rigs. It is also less vulnerable to sea-bed tremors, and if commuters get caught in a collapsed tunnel they will be beyond rescuing. A condition unlikely to happen to a bridge build in the open.

CONCERNING THE WORTHY SUGGESTIONS

For boosting our inventiveness, the following presumptions are imperative.

- Establishing the long awaited for **"The Invention Academy"**. (Page 41)

- There is superlative wisdom in summoning up our cognizants to collect and collate, **"The Encyclopedia Inventiliensis."** (Page 41)

- Establishing an independent, **"Governmental Commission"** to salvage the worthy inventions. (Page 457)

- Requesting our illustrious inventors to write about:

  - Corollary gadgetries that succeed a primal invention to enhance its proffer. (Page 39)

  - Demoded inventions condemned to oblivion. To mention a few: The Astrolabe, The Franklin Stove, The Onager, The Pantograph, The Steel Yard, The Universal Joint… etc. (Pages 39-40)

○   Inventions Conglomerates serving a common necessity and their differentiation into families and the families into lineages. (Page 17)

○   The monstrous Machines, from the stupendous moles to the one of a kind behemothic crawler capable of carrying the space shuttle to where the launching pad is. (Page 40)

•   Requesting our National Corporations to take part in:

○   Making **"The Tutorial Skull Cap"** since our know-how technologies are ripe for it. (Page 132) Involving the entire world, it will set up the stage for the open school venture. For this consignment, I choose Mr. Bill Gates. As a kudized habit of him, I am sure he will be using his proceedings to wipe the tears off the faces of the tormented by ignorance, hunger, disease, and homelessness.

○   Building the needed PLANTS together with the needed robotic machines, for tapping geothermal energy from the deeper hot layers of the earth's crust, to head off an expected shortage in fossil fuel. (Page 73)

Other suggestions that may seem peripheral to inventiveness, yet can eventuate beneficial ways for our harmonious co-existence, comprise:

- Coaxing all nations to incorporate their Green peace versions into their Parliamentary Systems. Hopefully, with all nations educating their peoples to conciliate with their environments, we may be able to extend this, "supposedly intelligent life of ours" into a brighter future. (Page 80)

- Enticing **ALL** nations to incorporate laws in their constitutions to stem the tide of hate crimes. (Page 151)

- Establishing a Global Moral Authority mustering an International Striking Force. Deployed, this force uses its leverage to separate and arbitrate between the warring factions. Yet if one turns to the worst, then this opponent is subdued to conform with the International Mandate. (Page 150)

- Making the zero calorie matrices from enzymatically destarched leguminous seeds to produce foods with different tastes and different caloric contents to help fighting obesity. (Page 50)

- Open not, Pandora's Box of cloning! Our natural resources and social services are in short supply. Hence we cannot afford any more on the Ark. (Page 154)

- Though a remote possibility, we should pre-notice our Mathematicians to develop, "The Numerical Language" needed to communicate with visitors from outer space. (Page 136)

- For appreciating difficulties of "Time Traveling", specially journeying into our past, requests were extended to a famous author and a famous director to produce a movie about "**The**

Big Crunch." In this movie "**The Book of Genesis**" is played in reverse order starting from our contemporaneous times to the point in which the **Supreme Creator** ordered nihility to split into matter and antimatter and into space and time. (Page 178 to 181)

The Beginning of the End

Not ready to accept my intuitions, sectors in our contemporaneous populace, were bent on ignoring my ideas and inventions. With this coming to pass, it was appropriate by me to trust my work to the coming generations.

Sure enough, there will be the appreciatives who will take my appealing ideas to completion. Also, there will be the insighted who, in their workshops, will take my peerless inventions from the drawing boards to materiality thence to applicability. To these ancillae, choosing to start from where I ended, I wish them the jubilances they deserve for daring to begin.

On the other hand, there are the objectionists, who chose to overlook my work. To these I say, the ideas I thought of, are here to deliver and to stay. Likewise, my inventions that I have done are here to do a world of good and to live after me. Assertedly, I know what I came up with, if not all being righteous-and who is perfect-most of it will be so.

My brain children, whether ideas or inventions, are my benefactions to America and the whole world.

APPENDIX IV.
FURTHER READINGS

Ackernecht, E.H., *Medicine at the Paris Hospital, 1794-1848.*John Hopkins University Press, 1967.

American Medical Association., *Family Medical Guide.*New York: Random House, Inc., 1987.

Attenborough, David., *The Living Planet.* Boston: Little, Brown and Company, 1984.

Attenborough, David., *The First Eden.* William Collins Sons & Co. Ltd., 1987.

Bilger, Burkhard et al., *Exploring Our World.* New York: Smith Mark Publisher, 1999.

Boyer, Carl B., *The History of Calculus and Its Conceptual Development.* New York: Dover Publications, 1949.

Briggs, Asa., *The Powers of Steam: An Illustrated History of the World's Steam Age.* Michael Joseph, 1982.

Burke, James., *The Day the Universe Changed.* Boston: Little, Brown and Company, 1985.

Burke, Peter., *Tradition and Innovation in Renaissance Italy.* Collins, 1972.

Clagett, Marshall., *History of Science.* University of Wisconsin Press, 1969.

Cooper, Brian, et al., *Multimedia, The Complete Guide*. New York: Dorling Kindersley Publishing, 1996.

Crowther, J.G., *Scientists of the Industrial Revolution*. Cresset Press, 1962.

Denton, Gillian, et al., *The History of the World*. New York: Dorling Kindersley Publishing, 1994.

Einstein, Elizabeth., *The Printing Press as an Agent of Change, Vol. I and II.,* Cambridge University Press, 1979.

Flinn, M.W., *Origins of the Industrial Revolution.*Longman, 1966.

Flower, Charles., *A Science Odyssey.* New York: William Morrow and Company Inc., 1998.

Grun, Bernard., *The Timetables of History, 3rd Edition.* New York: Simon and Schuster, 1991.

Hartwell, David G., *The Science Fiction Century*. New York: Tom Doherty Associates, Inc., 1997.

Heather, D.C., *Plate Tectonics*. Edward Arnold, 1979.

Hill, Richard., *Power in the Industrial Revolution*. Manchester: Manchester University Press, 1970.

Josephson, Mathew., *Edison: A Bibliography*. New York: McGraw-Hill, 1959.

Kerridge, Eric., *The Agricultural Revolution*. Manchester: Manchester University Press, 1970.

Kuhn, Thomas S., *The Structure of Scientific Revolutions*. Chicago: University of Chicago Press, 1962.

Lindberg, Davia C., *Theories of Vision from AL-Kindi to Kepler*. Chicago: University of Chicago Press, 1972.

Lorenz, Albert and Joy Schleh., *Ten Cities, Ten Centuries*. New York: Harry N. Abrahams, Inc., 1996.

Macauley, David., *The Way Things Work*. Boston: Houghton Mifflin Company, 1988.

Magee, Bryan., *The Story of Thought*. New York: Dorling Kindersley Publishing, 1998.

Meyer, Herbert W., *A History of Electricity and Magnetism*. Cambridge, Mass: MIT Press, 1971.

O'Neill, Richard., *Historical Facts: The Middle Ages*. New York: Crescent Books, 1992.

Parrinder, Geoffery., *World Religions: From Ancient History to the Present*. The Hamlyn Publishing Group, 1983.

Pledge, H.T., *Science Since 1500*. HMSO, 1966.

Rybczynski, Witold., *One Good Turn: A Natural History of the Screwdriver and The Screw.* New York: Scribner's, 2000.

Scholderer, Victor., *Johann Gutenberg: The Inventor of Printing.* British Museum Publications, 1970.

Stapleton, Michael., *The Illustrated Dictionary of Greek and Roman Mythology.* New York: Peter Bedrick Books, 1986.

Vess, D.M., *Medical Revolution in France.* University of Florida Press, 1975.

Watt, M., *Influence of Islam on Mediaeval Europe.* Edinburgh: Edinburgh University Press, 1972.

Weiner, Jonathan., *Planet Earth.* New York: Bantam Books, 1986.

Winston, Morton and Ralph Edelbach., *Society, Ethics and Technology.* Wadsworth Publishing Company, 1999.

Ziman, John., *Reliable Knowledge: Exploration of the Grounds for Belief in Science.* Cambridge University Press, 1978.

www.ingramcontent.com/pod-product-compliance
Lightning Source LLC
Chambersburg PA
CBHW031812170526
45157CB00001B/34